W9-CNC-284

DNA ARRAYS

TECHNOLOGIES AND EXPERIMENTAL STRATEGIES

METHODS & NEW FRONTIERS IN NEUROSCIENCE

Series Editors
Sidney A. Simon, Ph.D.
Miguel A.L. Nicolelis, M.D., Ph.D.

Published Titles

Apoptosis in Neurobiology
Yusuf A. Hannun, M.D., Professor/Biomedical Research and Department Chairman/ Biochemistry and Molecular Biology, Medical University of South Carolina
Rose-Mary Boustany, M.D., tenured Associate Professor/Pediatrics and Neurobiology, Duke University Medical Center

Methods for Neural Ensemble Recordings
Miguel A.L. Nicolelis, M.D., Ph.D., Professor/Department of Neurobiology, Duke University Medical Center

Methods of Behavioral Analysis in Neuroscience
Jerry J. Buccafusco, Ph.D., Professor/Pharmacology and Toxicology, Professor/Psychiatry and Health Behavior, Medical College of Georgia

Neural Prostheses for Restoration of Sensory and Motor Function
John K. Chapin, Ph.D., Department of Physiology, State University of New York Health Science Center
Karen A. Moxon, Ph.D., Department of Electrical and Computer Engineering, Drexel University

Computational Neuroscience: Realistic Modeling for Experimentalists
Eric DeSchutter, M.D., Ph.D., Department of Medicine, University of Antwerp

Methods in Pain Research
Lawrence Kruger, Ph.D., Professor Emeritus/Neurobiology, UCLA School of Medicine

Motor Neurobiology of the Spinal Cord
Timothy C. Cope, Ph.D., Department of Physiology, Emory University School of Medicine

Nicotinic Receptors in the Nervous System
Edward Levin, Ph.D., Associate Professor/Department of Pharmacology and Molecular Cancer Biology and Department of Psychiatry and Behavioral Sciences, Duke University School of Medicine

Methods in Genomic Neuroscience
Helmin R. Chin, Ph.D., NIMH, NIH Genetics Research
Steven O. Moldin, Ph.D, NIMH, NIH Genetics Research

Methods in Chemosensory Research
Sidney A. Simon, Ph.D., Professor/Department of Neurobiology, Duke University Medical Center
Miguel A.L. Nicolelis, M.D., Ph.D., Professor/Department of Neurobiology, Duke University Medical Center

The Somatosensory System: Deciphering the Brain's Own Body Image,
Randall Nelson, Ph.D., Professor of Anatomy and Neurobiology, University of Tennessee College of Medicine

New Concepts in Cerebral Ischemia
Rick Lin, Ph.D., Professor/Department of Anatomy, University of Mississippi Medical Center

DNA ARRAYS

TECHNOLOGIES AND EXPERIMENTAL STRATEGIES

Edited by Elena V. Grigorenko

CRC PRESS

Boca Raton London New York Washington, D.C.

Library of Congress Cataloging-in-Publication Data

DNA arrays : technologies and experimental strategies / edited by Elena V. Grigorenko.
p. cm. -- (Methods & new frontiers in neuroscience series)
Includes bibliographical references and index.
ISBN 0-8493-2285-5 (alk. paper)
1. DNA microarrays. I. Grigorenko, Elena V. II. Series.

QP624.5.D726 D624 2001
572.8'65--dc21 2001043455
CIP

Direct all inquiries to CRC Press LLC, 2000 N.W. Corporate Blvd., Boca Raton, Florida 33431.

Visit the CRC Press Web site at www.crcpress.com

International Standard Book Number 0-8493-2285-5
Library of Congress Card Number 2001043455
Printed in the United States of America 2 3 4 5 6 7 8 9 0
Printed on acid-free paper

Series Preface

Our goal in creating the *Methods & New Frontiers in Neuroscience Series* is to present the insights of experts on emerging experimental techniques and theoretical concepts that are, or will be, at the vanguard of neuroscience. Books in the series will cover topics ranging from methods to investigate apoptosis, to modern techniques for neural ensemble recordings in behaving animals. The series will also cover new and exciting multidisciplinary areas of brain research, such as computational neuroscience and neuroengineering, and will describe breakthroughs in classical fields like behavioral neuroscience. We want these books to be what every neuroscientist will use in order to get acquainted with new methodologies in brain research. These books can be given to graduate students and postdoctoral fellows when they are looking for guidance to start a new line of research.

The series will consist of case-bound books of approximately 250 pages. Each book will be edited by an expert and will consist of chapters written by the leaders in a particular field. The books will be richly illustrated and contain comprehensive bibliographies. Each chapter will provide substantial background material relevant to the particular subject. Hence, these are not going to be only "methods books." They will contain detailed "tricks of the trade" and information as to where these methods can be safely applied. In addition, they will include information about where to buy equipment, Web sites that will be helpful in solving both practical and theoretical problems, and special boxes in each chapter that will highlight topics that need to be emphasized along with relevant references.

We are working with these goals in mind and hope that as the volumes become available, the effort put in by us, the publisher, the book editors, and individual authors will contribute to the further development of brain research. The extent to which we achieve this goal will be determined by the utility of these books.

Sidney A. Simon, Ph.D.
Miguel A. L. Nicolelis, M.D., Ph.D.
Duke University
Series Editors

Preface

With advances in high-density DNA microarray technology, it has become possible to screen large numbers of genes to see whether or not they are active under various conditions. This is a gene-expression profiling approach that, over the past few years, has revolutionized the molecular biology field. The thinking is that any alterations in a physiological state are dictated by the expression of thousands of genes, and that microarray analysis allows that behavior to be revealed and to predict the clinical consequences. This rationale is sound enough, but until now it has not been substantiated by many experiments. The expectations for microarray technology are also high for prediction of better definition of patient groups, based on expression profiling. It is of obvious importance for assessing the efficacy of various treatments and to create "personalized" medicine.

The field of microarray technology presents a tremendous technical challenge for both academic institutions and industry. This book includes reviews of traditional nylon-based microarray assays as well as new, emerging technologies such as electrochemical detection of nucleic acid hybridization. Novel platforms such as oligonucleotide arrays are being developed, and companies that have never engaged in the life science industry are entering this rapidly growing market (see Dorris et al.'s review on oligonucleotide microarrays). Indeed, time will show which of the emerging technologies will have a significant impact on the future of microarray research.

Because microarray analysis is a high-throughput technology, the amount of data being generated is expanding at a tremendous rate. The handling and analysis of data require elaborate databases, query tools, and data visualization software. This book contains several examples of how a large set of data can be mined using different statistical tools (for details, see Chapters 6 and 7). Readers are also provided with a reproducible protocol for amplification of limited amounts of RNA in microarray-based analysis. The primary limitation of microrray technology — usage of a large amount of RNA — could be overcome with the technique described in Chapter 5 by Potier and colleagues, who in 1992 pioneered the RT-PCR technique for profiling gene expression in single neurons.

In summary, readers from different scientific fields and working environments will find this book a useful addition to the few books currently available. I am indebted to CRC Press Senior Editor Barbara Norwitz, who has given me unwavering support and brought common sense, order, and timeliness to a process that sometimes threatened to fall out of control. I also owe special thanks to Miguel Nicolelis for many good suggestions and Alexandre Kirillov for the encouragement and sustaining enthusiasm during the work on this book.

Editor

Elena V. Grigorenko, Ph.D., is a Scientist in the Technology Development Group at Millennium Pharmaceuticals, Inc., Cambridge, Massachusetts. She did her undergraduate studies in Russia at the Saratov State University and at the Moscow State University. Dr. Grigorenko's graduate research in bioenergetics was conducted in Dr. Maria N. Kondrashova's laboratory at the Institute of Biological Physics at Pushchino — a well-known biological center of the Russian Academy of Sciences. Dr. Grigorenko was a recipient of Sigma-Tau (Italy) and Chilton Foundation (Dallas, Texas) fellowships and she was a faculty member at the Wake Forest University School of Medicine, Winston-Salem, North Carolina. Currently her research interests are focused on applications of biochip and nanotechnologies for a drug discovery process.

Contributors

Bruno Cauli, Ph.D.
Neurobiologie et Diversité Cellulaire
ESPCI
Paris

Chris Clayton, Ph.D.
Glaxo Wellcome
Stevenage, U.K.

Sam A. Deadwyler, Ph.D.
Department of Physiology and Pharmacology
Wake Forest University School of Medicine
Winston-Salem, North Carolina

Frédéric Devaux, Ph.D.
Laboratorie de Génétique Moléculaire
ENS
Paris

David Dorris, Ph.D.
Motorola Life Sciences
Northbrook, Illinois

Allen Eckhardt, Ph.D.
Xanthon, Inc.
Research Triangle Park, North Carolina

Holger Eickhoff, Ph.D.
Max-Planck-Institut für Molekulare Genetik
Berlin

Eric Espenhahn, Ph.D.
Xanthon, Inc.
Research Triangle Park, North Carolina

Willard M. Freeman, Ph.D.
Department of Physiology and Pharmacology
Wake Forest University School of Medicine
Winston-Salem, North Carolina

Stefanie Fuhrman, Ph.D.
Incyte Genomics, Inc.
Palo Alto, California

Alexander Gee, Ph.D.
AnVil Informatics, Inc.
Lowell, Massachusetts

Natalie Gibelin, Ph.D.
Neurobiologie et Diversité Cellulaire
ESPCI
Paris

Geoffroy Golfier, Ph.D.
Neurobiologie et Diversité Cellulaire
ESPCI
Paris

Elena V. Grigorenko, Ph.D.
Millennium Pharmaceuticals, Inc.
Cambridge, Massachusetts

Georges Grinstein, Ph.D.
AnVil Informatics, Inc.
Lowell, Massachusetts

Bruce Hoff, Ph.D.
BioDiscovery, Inc.
Los Angeles, California

Patrick Hoffman, Ph.D.
AnVil Informatics, Inc.
Lowell, Massachusetts

C. Bret Jessee, Ph.D.
AnVil Informatics, Inc.
Lowell, Massachusetts

Josef Kittler, Ph.D.
University College of London
London

Sonia Kuhlmann, Ph.D.
Neurobiologie et Diversité Cellulaire
ESPCI
Paris

Alexander Kuklin, Ph.D.
BioDiscovery, Inc.
Los Angeles, California

Bertrand Lambolez
Neurobiologie et Diversité Cellulaire
ESPCI
Paris

Beatrice Le Bourdelles
Neuroscience Research Centre
Merck Sharp & Dohme Research Laboratories
Harlow, United Kingdom

Hans Lehrach, Ph.D.
Max-Planck-Institut für Molekulare Genetik
Berlin

Shoudan Liang
Incyte Genomics, Inc.
Palo Alto, California

Scott Magnuson, Ph.D.
Motorola Life Sciences
Northbrook, Illinois

Philippe Marc
Laboratorie de Génétique Moléculaire
ENS
Paris

Abhijit Mazumder, Ph.D.
Motorola Life Sciences
Northbrook, Illinois

Mary Napier, Ph.D.
Xanthon, Inc.
Research Triangle Park, North Carolina

Wilfried Nietfeld, Ph.D.
Max-Planck-Institut für Molekulare Genetik
Berlin

Eckhard Nordhoff, Ph.D.
Max-Planck-Institut für Molekulare Genetik
Berlin

Lajos Nyarsik, Ph.D.
Max-Planck-Institut für Molekulare Genetik
Berlin

Phil O'Neil, Ph.D.
AnVil Informatics, Inc.
Lowell, Massachusetts

Natasha Popovich, Ph.D.
Xanthon, Inc.
Research Triangle Park, North Carolina

Marie-Claude Potier, Ph.D.
Neurobiologie et Diversité Cellulaire
ESPCI
Paris

Ramesh Ramakrishnan, Ph.D.
Motorola Life Sciences
Northbrook, Illinois

Jean Rossier, Ph.D.
Neurobiologie et Diversité Cellulaire
ESPCI
Paris

Ulrich Schneider, Ph.D.
Max-Planck-Institut für Molekulare Genetik
Berlin

Tim Sendera
Motorola Life Sciences
Northbrook, Illinois

Shishir Shah, Ph.D.
BioDiscovery, Inc.
Los Angeles, California

Soheil Shams, Ph.D.
BioDiscovery, Inc.
Los Angeles, California

Roland Somogyi, Ph.D.
Molecular Mining Corporation
Kingston, Ontario, Canada

Holden Thorp, Ph.D.
Department of Chemistry
Kenan Laboratories
University of North Carolina at Chapel Hill
Chapel Hill, North Carolina

Kent E. Vrana, Ph.D.
Department of Physiology and Pharmacology
Wake Forest University School of Medicine
Winston-Salem, North Carolina

Don Wallace, Ph.D.
Glaxo Wellcome
Stevenage, U.K.

Xiling Wen, Ph.D.
Incyte Genomics, Inc.
Palo Alto, California

Robert Witwer, Ph.D.
Xanthon, Inc.
Research Triangle Park, North Carolina

Günther Zehetner, Ph.D.
German Resource Centre and Primary Database in the German Genome Project
Berlin

Shou-Yuan Zhuang, Ph.D.
Department of Physiology and Pharmacology
Wake Forest University School of Medicine
Winston-Salem, North Carolina

Contents

1 Technology Development for DNA Chips

Holger Eickhoff, Ulrich Schneider, Eckhard Nordhoff, Lajos Nyarsik, Günther Zehetner, Wilfried Nietfeld, and Hans Lehrach

CONTENTS

1.1 DNA MICROARRAYS: METHOD DEVELOPMENT

The identification of the DNA structure as a double-stranded helix consisting of two nucleotide chain molecules was a milestone in modern molecular biology. Most of the methods for DNA characterization are based on its ability to form fully or partially complementary double helices from two complementary single strands. To detect hybridization events, one strand (target) is usually immobilized on a solid support (e.g., nylon membranes or glass slides), whereas its counterpart (probe) is present in the hybridization solution. The probe is labeled and hybridization events are thereby detected on the solid support at the position of the immobilized target. Hybridization with different known probes can be used to characterize unknown targets, such as is used in oligonucleotide fingerprinting. The reverse situation — the target DNA is known and the hybridization solution is not defined — is encountered when DNA chips or microarrays are used to monitor gene expression.

0-8493-2285-5/02/$0.00+$1.50

The automated procedures established in this and other laboratories include the following steps: clone picking, clone spotting, hybridization, detection, image analysis, and computer analysis, including primary data storage of hybridization event.[1] For high-throughput DNA analyses, DNA molecules are randomly fragmented and then introduced into the bacterial plasmids. Colonies of transformed bacteria are grown on agar plates such that each colony carries a single DNA fragment (clone). The entirety of these clones forms a clone library. Each carries a relatively short DNA fragment, between 100 and 4000 bp in length. A large number of clones must be provided for full coverage of a genome or a tissue-specific library. A typical tissue-specific library consists of a few hundred thousand clones. Selected clones are picked, propagated, and stored in 384-well microtiter plates. This allows long-term storage, analysis, and subsequent individual clone retrieval. Clones from microtiter plates can be used for DNA amplification by PCR, spotted on a surface, and hybridized with specific or complex probes.[2,3]

The first generation of clone picking and spotting robots with stepper motors was invented between 1987 and 1991 in the laboratory of Hans Lehrach at the Imperial Cancer Research Fund in London.[4,5] The XYZ systems at that stage were purchased from Unimatic Engineers Ltd., London, and from the former ISERT Electronics, Eitersfeld, Germany, which is now called ISEL Automation. These first-generation machines, using two-phase stepper motors from Orientel or Vextar in a half-step modus with 400 steps/rotation, achieved a 1/100-mm resolution. The robots have been programmed for a spatial resolution of 0.015 mm over a moving length of 600 mm (39,000 steps in the x direction, resp. 38,000 in the y direction). These instruments achieved spotting densities of more than 400 dots/cm^2.

More powerful spotting devices were engineered during the years 1991–1992 and implemented in second-generation robots, which utilized linear motors and were equipped with blunt-end and split pins for DNA transfer onto nylon and glass (see Figure 1.1).[6] The original motors were purchased from Linear Technology Ltd., now called Linear Drives Ltd. These robots had a much wider movement range (approximately 1000 × 750 × 150 mm). The package utilized special INA bearings, 0.2-mm encoders, LTL drives, as well as control electronics programmed over an RS232

FIGURE 1.1 Liquid delivery onto a slide surface with a solid pin. From left to right, the four-picture sequence illustrates the liquid delivery onto an epoxysilanized surface with a 250-μm pin. The amount of liquid in the droplet was measured to 2 nl. From the sequence it is clear that not the whole droplet is transferred to the slide because approximately 5% of the droplet splashes back to the print tip's end.

serial port. In addition, devices for plate handling and temporary removal of microtiter-plate lids were implemented. The instrumentation was able to spot up to 2500 dots/cm^2. Today, upgraded versions of these machines are in use in many laboratories. They have been further improved, mainly by the integration of more accurate and faster drives combined with better encoders, providing higher sample throughput and superior reproducibility.

1.2 EVOLUTION OF THE PIN DESIGN

The transfer of clones and PCR products was first achieved with solid pins, manufactured from stainless steel (see Figure 1.1). These pins had a print tip 0.9 mm in diameter. Many different shapes of solid pins have been manufactured and tested for optimal transfer of the target DNA onto the support. Current solid pins with either a conical or cylindrical print tip have diameters down to 100 μm. Different support materials have been tested, including titanium, tungsten, and mixtures thereof. An important advantage of solid pins is that they can be easily cleaned and sterilized. For this purpose, they are usually flushed in a bath containing bleaching agents and an upside-down brush. Over the past years, it has been shown that these pins can perform thousands of sample transfers without loss of spotting performance. A major disadvantage of solid pins is the fact that after one loading procedure, only one slide or filter can be addressed for spotting. This is especially time-consuming when the same spot on the planar surface must be addressed several times in order to deposit sufficient DNA material for hybridization purposes, or when a large number of array replicates must be produced. This limitation was overcome by designing split pins that can accommodate up to 5 μl liquid by capillary forces. These pins allow spotting of more than ten glass slides before the pins have to aspirate liquid the next time. Compared to linear solid pins, split pins are more difficult to clean, and the production costs are up to 100 times higher. The volume delivered with both pins is in the range 0.5 to 5 nl, primarily determined by the print tip diameter or the dimensions of the enclosed cavity (split pins).

As an alternative to conventional needle spotting technology, a drop-on-demand technology was developed. To reduce the dimension of arrays by one or two orders of magnitude, the samples are now pipetted with a multichannel microdispensing robot.[7] The principle is similar to that of an inkjet printer. A two-dimensional, 16-nozzle head is moved in *x*, *y*, and *z* directions with 5-μm resolution using a servo-controlled linear drive system (see Figure 1.2). The spacing between the dispenser capillaries enables the aspiration of samples provided in microtiter plates of different formats. After aspirating the samples, each nozzle moves to a different drop inspection system. Integrated image analysis routines decide whether or not a suitable drop is generated. If the drop is poorly formed, automated procedures clean the nozzle tip. A second integrated camera defines the positions for automated dispensing (e.g., filling of cavities in silicon wafers). Each nozzle is able to dispense single or multiple drops with a volume of 100 pl. We recently introduced a magnetic bead-based purification system inside the dispensers. This allows concentration and purification prior to dispensing. The resulting spot size depends on the surface and varies between 100 and 120 μm. The density of the arrays can be increased to 3000 spots/cm^2. The

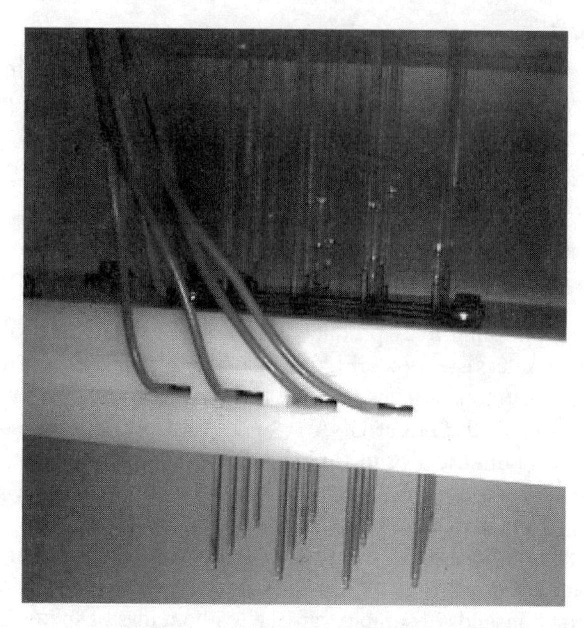

FIGURE 1.2 Two-dimensional piezo ink jet arrayer. In this 16-nozzle arrayer, each of the 16 jets can aspirate and dispense individually. The device is mounted into a cartesian robot system and delivers 80-pl droplets on-the-fly onto up to 80 slides in parallel. The nozzles are mounted in a spacing that allows for aspiration and dispensing from 1536 well plates.

functionality of the microdispensing system allows one to dispense on-the-fly and it takes less than 3 minutes to put 100 × 100 spots in a square, each spot being 100 µm in diameter and the distance between the centers of two spots being 230 µm. At this density, it is possible to immobilize a small cDNA library consisting of 14,000 clones on the surface of one microscope slide. This offers a higher degree of automation because glass slides are easier to handle than nylon membranes.

1.3 EVOLUTION OF THE DNA CARRIERS OR SUPPORTS

DNA arrays on nylon membranes are a widely used tool in modern molecular biology. The founding of the Reference Library System in 1992 was the first step in providing scientists who do not own an arrayer with clones in an ordered format. Nitrocellulose and nylon membranes containing up to 100,000 DNA fragments per 22 cm × 22 cm membrane show good DNA binding capacity and offer the possibility of reusing the arrays up to ten times. Although working reliably in many laboratories, alternatives to nylon membranes were sought because most nylon membranes display an inherent fluorescence signal, which prohibits all fluorescence-based detection methods. Although it was shown that single clones can be identified on nylon filters with enzyme-amplified fluorescence, the background on nylon membranes for non-amplified signals, as required for quantitative hybridization assays, stayed much too high.

To meet these requirements, attachment procedures for the immobilization of DNA on glass were developed. At present, two main strategies are followed for the DNA immobilization on glass. They are based either on covalent attachment procedures or hydrophobic interactions. One important feature for all noncovalent DNA immobilization methods is the hydrophobicity of the coating on the glass slide. A useful test whether or not a polylysine slide is ready for spotting is the 45° lifting of one slide corner. A predeposited 1-μl water droplet must move without a smear to obtain good spotting results (Brown, P.O., personal communication). For the majority of covalent attachment procedures, the PCR product is often modified with primers carrying 5′ amino groups, which allow fixture to amino-derivatized glass via dialdehydes or directly to epoxysilanated glass slides. Although the scheme looks quite simple, a number of parameters, such as linker length either on the PCR product or on the surface, play an important role for maximum binding and hybridization efficiencies.[8]

As a result of the mainly two-dimensional structure on the glass surface and independent of the immobilization procedure, only 10% of the DNA can be immobilized on a specified glass area when compared to the fibrillic, three-dimensional structure of nylon membranes. This results in very tiny amounts of DNA on the slide, which require very sensitive detection devices. An optimized and modified planar surface produces a three-dimensional structure on a glass slide through a chemistry that creates a dendritic structure of polymers.[9]

New developments for the improvement of filter technology include their lamination onto plastics[10] and glass slides to enable better handling with increased binding properties (Schleicher and Schüll, Dassel, Germany). Preliminary results show the suitability of these low-fluorescence background materials for fluorescence-based quantitative hybridization assays.

Gel-based arrays might the optimal surface for protein arrays because proteins need a nearly physiological environment to stay in their native folding. This can be achieved in gel matrices on glass slides,[11] which present a further development of currently used membranes.[12] Currently, a number of researchers are investigating whether polished and therefore very flat (superflat) glass slide surfaces with a height deviation of at most 1 μm can improve the accuracy of the results. Although likely, the results published thus far are insufficient to draw solid conclusions (http://www.arrayit.com).

1.4 LABELING

Over the past 7 years fluorescent labeling technologies have accompanied the increased usage of glass slide-based DNA chip technology. Although different incorporation rates of either Cy3- or Cy5-labeled triphosphates during reverse transcription might cause uncertainties in the linear performance of the two dyes over the detection range, they are widely used.[13]

Alternatives to direct fluorescence-based detection are enzyme-amplified fluorescence,[3] radioactivity-based,[14,15] and mass spectrometric detection methods.[16] The main disadvantage of monocolor detection methods, when compared to the simultaneous detection of two fluorescent dyes, is that the use of chemoluminescence or radioactive labels requires two separate hybridization experiments to compare two different expression profiles. In addition to health considerations, the use of radioactive labels such as ^{32}P, ^{33}P, or ^{35}S at high sample density suffers from the direction-independent emission, yielding diffuse signals on the autoradiographs. Nevertheless, we have observed that radioactive detection on glass slides provides at least a fivefold increased sensitivity in expression profiling experiments when compared to fluorescence.[17]

An alternative for label-free DNA hybridization detection might be a detection scheme that uses mass spectrometry. Mass spectrometry separates molecular ions according to their charge-to-mass ratio prior to the detection, which opens up higher-order multiplexing than is possible using the different fluorescent dyes. The detection of DNA at high resolution, however, is currently limited to <100 nucleotides.[18] The detection sensitivity lags more than three orders of magnitude behind fluorescence-based detection methods, and the analysis is considerably more time-consuming. Due to these limitations, mass spectrometry for gene expression profiling is not currently an attractive alternative to fluorescence-based detection systems. For other, equally important applications of DNA chip technology, such as the detection of single nucleotide polymorphisms, MALDI-MS has proven to be very efficient.[16] Compared to expression profiling, the molecules being detected are significantly smaller. These can be short oligonucleotides generated in a primer extension reaction, by the invader assay, or short hybridized PNA oligomers. In all cases, compared to, for example, DNA >50 nucleotides, both the detection sensitivity and the signal resolution are considerably higher. The latter allows efficient multiplexing. While radiolabeling methods clearly dominated biotechnology in the past, light-optical principles and mass spectrometric detection methods will dominate DNA chip technology in the near future.

1.5 HYBRIDIZATION

In the past 10 years, hybridization experiments using nylon filters were either performed in polyethylene bags or in roller bottles inside hybridization ovens. The majority of protocols published for glass slide hybridizations is such that 10 μl of hybridization solution containing the probe is transferred to the microarray and covered with a coverslip, which forms a thin probe film. This setup is then incubated at 42°C in a humidity chamber. After incubation (e.g., overnight for expression analysis based on 1.0 μg of poly-RNA), the arrays are washed and scanned.

We have developed the slide sandwich principle (SSP), in which the coverslip is replaced by another slide. Therefore, two spotted microarrays when placed face to face are incubated with the same probe solution. The technology is independent of glass slide size and has been tested for slides up to an area of 8 cm × 12 cm. One basic advantage of the SSP is that two data sets deriving from one probe can be scored in one experiment. In another setup, we replaced the normal coverslip with a quartz double-bandpass filter containing inlet and outlet valves for liquid handling, and mounted into a peltier thermostatic holder. This setup allows monitoring and final detection of fluorescent-labeled hybridization probes online.

1.6 OUTLOOK AND CHALLENGES

Combining the disciplines of microfabrication, chemistry, and molecular biology is a promising approach for future developments. We will witness the development of chip biology, which adopts methodology, management, and technology related to the semiconductor industry. A prominent example of this is the generation of high-density probe arrays by on-chip, solid-phase oligonucleotide synthesis controlled by light and the use of photolithographic masks. The high-throughput screening methods would benefit from further automation and miniaturization. Along with the ongoing miniaturization process in biotechnology, new hardware tools will have to be developed. In addition to all the necessary handling steps required for on-chip hybridization experiments, the existing detection systems in particular need to be improved. Lower spot sizes require more sensitive detection systems, which puts stress on the spatial resolution power.

Another prerequisite for further improvements in DNA chip technology is the introduction of cleanroom facilities in modern molecular biology laboratories. As in the semiconductor industry, dust, dandruff, and other microparticles disturb the manufacturing process (sticking to pins, clogging dispenser nozzles, producing false positive signals). In addition, the use of manufactured chips also requires a clean environment. For example, microparticles can block hybridization events on the chip surface or produce false positive signals during the analysis of the chip.

The introduction of commonly accepted quality controls that allow for comparing the results produced in different laboratories is another requirement for future development. The Max Planck Institute (MPI) for Molecular Genetics has proposed to include at least two controls in all experimental setups. For all applications in mammalian systems, we use plant-specific genes that are spotted into every spotting block as a dilution series. For plant-specific chips, we have chosen the opposite approach and selected two mammalian-specific control clones. All applications have in common that one control clone is spiked into the labeling reaction while the other one is labeled in a separate container. Both reactions are combined and exposed to the microarray simultaneously. This procedure allows one to normalize the data retrieved from microarrays for different labeling yields, hybridization efficiencies, and sample spotting deviations. The dilution series within the control clones allows one to determine the dynamic range for a specific experiment (Schuchardt, J. et al., *Nucleic Acids Research*, in press). Said control clones can be obtained via the Resource Centre in the German Genome Project (http://www.rzpd.de).

Together with all the technical developments, the success of DNA microarrays will greatly depend on the bioinformatic tools available. Bioinformatics in the DNA microarray field starts with fully automated and batchwise working image analysis programs and should cover all aspects of statistical analyses (reproducibility of experiments, background determination, clustering, etc.) and their link to gene regulation and function. The graphical DNA Array Displayer developed jointly by the MPI for Molecular Genetics and the Resource Centre within the German Genome Project covers some aspects of these requirements. The Displayer allows one to track all the information about previous experiments available for each clone that is present in a particular array.

ACKNOWLEDGMENTS

The authors would like to thank the Bundesministerium für Bildung und Forschung for its financial support within the projects "Automation in Genome Analysis" and "Slide."

REFERENCES

1. Maier, E., Robotic technology in library screening, *Laboratory Robotics and Automation*, 7, 123–132, 1995.
2. Gress, T. M., Hoheisel, J. D., Lennon, G. G., Zehetner, G., and Lehrach, H., Hybridization fingerprinting of high-density cDNA-library arrays with cDNA pools derived from whole tissues, *Mammalian Genome*, 3, 609–619, 1992.
3. Maier, E., Crollius, H., and Lehrach, H., Hybridization techniques on gridded high density DNA *in situ* colony filters based on fluorescence detection, *Nucl. Acids Res.*, 22, 3423–3424, 1994.
4. Poustka, A., Pohl, T., Barlow, D. P., Zehetner, G., Craig, A., Michiels, F., Ehrich, E., Frischauf, A. M., and Lehrach, H., Molecular approaches to mammalian genetics, *Cold Spring Harbor Symposia on Quant. Biol.*, 51, 131–139, 1986.
5. Lehrach, H., Drmanac, R., Hoheisel, J., Larin, Z., Lennon, G., Monaco, A.P., Nizetic, D., Zehetner, G., and Poustka, A., Hybridization fingerprinting in genome mapping and sequencing, in *Genome Analysis, Vol. 1: Genetic and Physical Mapping*, Cold Spring Harbor Laboratory Press, Cold Spring Harbor, New York, 1990, 39–81.
6. Lennon, G. G. and Lehrach, H., Hybridization analyses of arrayed cDNA libraries, *Trends Genet.*, 7, 314–317, 1991.
7. Eickhoff, H., Microtechnologies and miniaturization, *Drug Discovery Today*, 3, 148–149, 1998.
8. Graves, D. J., Su, H. J., McKenzie, S. E., Surrey, S., and Fortina, P., System for preparing microhybridization arrays on glass slides, *Anal. Chem.*, 70, 5085–5092, 1998.
9. Matysiak, S., Hauser, N., Wurtz, S., and Hoheisel, J., Improved solid supports and spacer/linker systems for the synthesis of spatially addressable PNA-libraries, *Nucleosides Nucleotides*, 18, 1289–1291, 1999.
10. Bancroft, D., Obrien, J., Guerasimova, A., and Lehrach, H., Simplified handling of high-density genetic filters using rigid plastic laminates, *Nucl. Acids Res.*, 25, 4160–4161, 1997.
11. Arenkov, P., Kukhtin, A., Gemmell, A., Voloshchuk, S., Chupeeva, V., and Mirzabekov, A., Protein microchips: use for immunoassay and enzymatic reactions, *Anal. Biochem.*, 278, 123–131, 2000.
12. Lueking, A., Horn, M., Eickhoff, H., Bussow, K., Lehrach, H., and Walter, G., Protein microarrays for gene expression and antibody screening, *Anal. Biochem.*, 270, 103–111, 1999.
13. Iyer, V. R., Eisen, M. B., Ross, D. T., Schuler, G., Moore, T., Lee, J. C. F., Trent, J. M., Staudt, L. M., Hudson, J., Boguski, M. S., Lashkari, D., Shalon, D., Botstein, D., and Brown, P. O., The transcriptional program in the response of human fibroblasts to serum, *Science*, 283, 83–87, 1999.
14. Nguyen, C., Rocha, D., Granjeaud, S., Baldit, M., Bernard, K., Naquet, P., and Jordan, B. R., Differential gene expression in the murine thymus assayed by quantitative hybridization of arrayed cDNA clones, *Genomics*, 29, 207–216, 1995.

15. Granjeaud, S., Nguyen, C., Rocha, D., Luton, R., and Jordan, B. R., From hybridization image to numerical values: a practical, high throughput quantification system for high density filter hybridizations, *Genetic Anal.*, 12, 151–162, 1996.
16. Griffin, T. and Smith, L. M., Single-nucleotide polymorphism analysis by MALDI-TOF mass spectrometry, *Trends Biotechnol.*, 18, 77–84, 2000.
17. Maier, E., Meierewert, S., Bancroft, D., and Lehrach, H., Automated array technologies for gene expression profiling, *Drug Discovery Today*, 2, 315–324, 1997.
18. Stomakhin, A., Vasiliskov, V. A., Tomofeev, E., Schulga, D., Cotter, R., and Mirzabekov, A., DNA sequence analysis by hybridization with oligonucleotide microchips: MALDI mass spectrometry identification of 5mers contiguously stacked to microchip oligonucleotides, *Nucl. Acids Res.*, 28, 1193–1198, 2000.

2 Experimental Design for Hybridization Array Analysis of Gene Expression

Willard M. Freeman and Kent E. Vrana

CONTENTS

0-8493-2285-5/02/$0.00+$1.50

2.1 INTRODUCTION

Given the explosion in genomic information, the historical "one gene at a time" approach to gene expression analysis is no longer adequate. Instead, large-scale multiplex methods for analyzing gene expression patterns are needed. Several technologies have been developed to serve this function, including differential display, serial analysis of gene expression (SAGE), total gene expression analysis (TOGA), subtraction cloning, and DNA hybridization arrays (microarrays).[1] This last approach, which is rapidly becoming the dominant technology in the gene expression field, is the subject of this chapter. However, this powerful new technology also comes with a unique set of considerations when it comes to designing and executing experiments. In this chapter, experimental design will be considered from both strategic and tactical standpoints.

In the three decades since the first recombinant DNA technologies were introduced, the standard paradigm has been to examine and characterize the sequence and expression of one or two genes at a time. At best, this approach involved the time- and labor-intensive sequential analysis of gene products in a given pathway. At worst, in the case of complex polygenic phenotypes or diseases, this time-consuming process has severely limited the ability of the molecular biology research community to move scientific understanding forward. The vast amounts of genomic data being generated by the Human Genome Project are exacerbating this problem. In June 2000, researchers announced the completion of a rough draft of the human genome — the beginning of what some are the calling the *postgenomic era* (a period of research in which the question is not how to sequence the genome, but what to do with the complete sequence). By 2001/2002, a high-fidelity sequence for all human genetic material will be available, providing detailed information on the estimated 100,000 genes required to encode a human being. In this postgenomic era of research, the old practices of "one gene at a time" will be inefficient and unproductive. Such approaches would not only be inefficient but would not sufficiently illuminate *patterns* of gene expression; therefore, they will be inappropriate for analyzing complex diseases or physiological/behavioral/pharmacological states.

2.2 ROLE OF HYBRIDIZATION ARRAYS IN FUNCTIONAL GENOMICS

The current challenge, therefore, is to develop/optimize methods for monitoring thousands of gene products simultaneously (genomic-scale analysis of gene expression). To this end, functional genomics is becoming a dominant feature of the molecular biology landscape (Figure 2.1 shows the various types of genetic information that can be mined). For the purpose of this chapter, functional genomics is defined as the study of all the genes expressed by a specific cell or group of cells and the changes in their expression pattern during development, disease, or environmental exposure. DNA polymorphism analysis is sometimes included under functional genomics, but for this chapter it is included under genomics. With this definition in mind, we can say that functional genomics is simply large-scale gene expression analysis at the RNA level. Given that each cell in an organism inherits

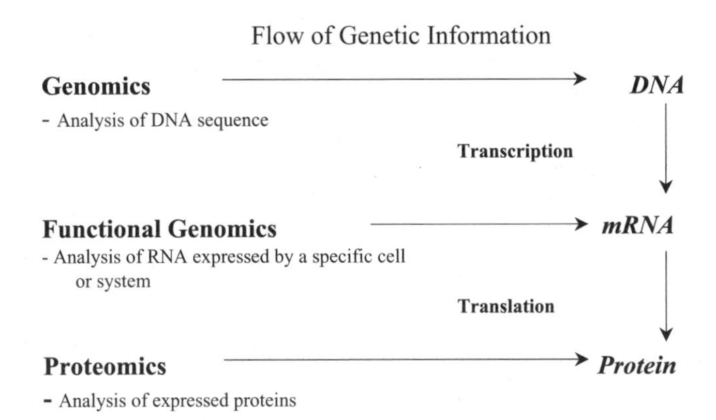

FIGURE 2.1 Genetic information flows from DNA into mRNA through transcription and then from mRNA to protein through translation. It should be noted that there is some controversy over whether polymorphism analysis should be included in functional genomics. For the present discussion, we chose to include this under genomics because it represents structural variations in DNA sequence — albeit with the potential to represent functional changes.

a constant genetic legacy (the DNA contained within the nucleus), it is the *pattern* of specific genes that is expressed that establishes the identity of a given cell or tissue. Analysis of these patterns in the context of the administration of drugs, in various disease processes, or following exposure to toxins, will be central to understanding biology and how humans respond, on a molecular level, to these conditions.

Biological research and discovery in the postgenomic era will require management of an incredible wealth of information. The question is no longer one of being able to sequence genomes but what to do with the sequences. The vast amount of genetic information being generated by sequencing projects will not only tax our existing methods of data collection and management but will require us to change our fundamental experimental mind-set. We will no longer be interested in individual genes; rather, the emphasis will be on the analysis of patterns of gene expression.

Returning to Figure 2.1, note that molecular-biological analysis can occur at three different levels. Most of the previous work has focused on the genomic, or DNA, level. Diseases have traditionally been examined by mapping inherited disorders with traditional genetic methods. Alternatively, individual genes were cloned (based on rational biochemical insights) and characterized relative to a disease or physiological response. Now, a new generation of genomic technologies will take the dominant position. These technologies allow rapid sequencing of DNA for diagnostic and research purposes and genome scans for single nucleotide polymorphisms (SNPs). SNPs are single base-pair variations in DNA that may cause disease or be useful as markers of disease. While extremely important, work at the DNA level does not answer all questions associated with the transcription of RNA and the translation of protein — gene expression. For example, exposure to a neurotoxin may induce the expression of a programmed cell death (apoptosis) pathway, leading to neurodegeneration. Such a change in gene expression in response to an environmental insult might be unrelated to a specific sequence polymorphism and yet still represent a valuable therapeutic target for drug design. None of the traditional genomic approaches — nor most of the new SNP analysis methods — is well suited to broad-based gene expression studies.

One of the best ways (if not theoretically *the* best way) to study gene expression is to examine the proteins encoded by genes. Studying all the proteins expressed in a cell is known as *proteomics*.[2] By comparing protein patterns in treated vs. untreated tissues or in diseased vs. nondiseased tissues or cells, researchers can pinpoint the proteins involved in disease processes, proteins that could be targets of novel therapies. Proteins, after all, are the key to realizing the potential encoded in the genome. Unfortunately, proteomic analysis, although clearly the best choice, is technically tedious (involving two-dimensional protein electrophoresis), requires sophisticated infrastructure (mass spectrometry), and is not necessarily high-throughput in nature. These characteristics have placed this approach beyond the reach of most investigators outside of the large pharmaceutical companies and have made companies that have improved the technology unwilling to publicize their progress for proprietary reasons.

The other means of gene expression analysis is functional genomics, which, on the surface, is not the stage-of-choice for analyzing gene expression because RNA is a transitional step from DNA to protein. Indeed, RNA has limited value except as a protein precursor. However, functional genomics can build upon the base of knowledge generated by the Human Genome Project to simultaneously examine the expression of thousands of genes. This large-scale expression analysis is possible because gene-specific probes for mRNA can be generated from DNA sequence information. Once identified at the level of mRNA, alterations in gene expression can be extended to protein. The functional genomic analysis therefore helps to identify target proteins for additional study.

The limitations of examining mRNA levels are that it does not provide direct insight into underlying polymorphisms (SNPs) that could be basis of disease, and that just because an mRNA level changes does not mean the corresponding protein levels must change.[3] In addition, mRNA measurements do not account for changes that a protein may undergo (glycosylation, phosphorylation, subcellular targeting, etc.) after it is produced. However, hybridization array technology is readily available and can be accessed by nearly any laboratory to provide valuable insights into functional genomics. The key point is that there are unique problems associated with this technology that must be taken into account.

2.3 STRATEGIC CONSIDERATIONS IN ARRAY EXPERIMENTAL DESIGN

The main reason for undertaking DNA hybridization analysis is to accomplish two important goals. The first is to provide a broad-based screen of gene expression. The desire is to effectively and economically filter through thousands of genes to identify those that are regulated by a physiological or pharmacological intervention. As the field rapidly accumulates knowledge on the 100,000 or so distinct genes, this will prove to be the only way to effectively study biological processes. A second goal is to actually understand *patterns* of gene expression. We will soon be in a position to understand not only how genes are regulated in isolation, but how families of genes or members of common regulatory pathways are coordinately regulated. Therefore, the strategic implications of how we recognize and analyze patterns of gene expression will be at least as important as the array technology itself.

2.3.1 Large-Scale Functional Genomic Screening

Initial functional genomic screens seek to establish what genes are expressed in a given cellular population and what genes appear to be regulated by experimental conditions as compared to control conditions. Large-scale screens are initially needed because the full complement of genes expressed in different tissues and cells is usually unknown. While much may be known about the genes expressed in a particular cell, this set of genes may change under the experimental conditions. Although the genes contained on the arrays used for this initial screen can be very large, the array will most likely be incomplete. The overriding principle of this step in the process is "hypothesis generation."[4] That is, large-scale DNA arrays should be considered a means for creating testable hypotheses.

There are three main platforms available for large-scale gene expression scans: macroarrays, microarrays, and high-density oligonucleotide arrays. The nomenclature of the field sometimes uses these terms interchangeably; but for the purposes of this discussion, these terms refer to specific types of hybridization arrays.[5] Macroarrays use a membrane array matrix, radioactively labeled targets for detection, and the samples are hybridized to separate arrays. This form of array generally contains between 1000 and 10,000 genes. Several different arrays can be used to give even broader coverage. Microarrays use a glass or plastic matrix with fluorogenically labeled targets, and the targets are competitively hybridized to the same array. These arrays can contain up to tens of thousands of genes. Finally, high-density oligonucleotide chips use *in situ* constructed olgonucleotides for probes. Samples are hybridized to separate arrays and a fluoroprobe is used for detection. These arrays also contain up to tens of thousands of genes. Each of these formats has different advantages and limitations in terms of number of genes, model organisms available, sensitivity, and cost.

2.3.2 *Post hoc* Confirmation of Changes

Post hoc confirmation is a critical step in functional genomic research and yet it is often underrepresented in the literature. While initial large-scale screening can produce a number of targets, that screen is not the final experiment. The targets generated from the large-scale screening are like suspects in a police lineup, and the *post hoc* confirmation is the beginning of proving a scientific case for which gene(s) are responsible for the biological phenomenon being studied. Confirmation can be achieved at the level of nucleic acids (Northern blotting or QRT-PCR[6]) or at the level of protein (immunoblotting and other proteomic approaches). These are discussed further in Section 2.4.3.

2.3.3 Custom Arrays

Custom arrays serve as a form of hypothesis-testing in functional genomic experiments. These arrays contain a smaller set of genes than the large-scale screening arrays and are focused on genes and gene families highlighted in large-scale screens. The advantage of custom arrays is that they can exhaustively examine a smaller set of genes. This is an advantage, both scientifically and practically. Because large arrays often contain only a few members/isoforms of specific gene families, custom arrays can be constructed that contain all of the subtypes and splice variants. As well, the cost of custom hybridization arrays is often less when measured on a per-gene basis.

There are a number of technical considerations with generating custom arrays.[7,8] The key is in selecting the probes placed on the array. Probes must be carefully designed to discriminate between highly homologous genes. In addition, multiple spots of the same gene per array increase confidence intervals. Finally, with the low cost per custom array (after initial start-up), more replicates of the experiment can be performed, and arrays can be applied to individual animals/samples. All of these steps combine to allow detailed investigation of the hypothesis generated from the initial large-scale screen and *post hoc* confirmation.

2.3.4 Bioinformatics

Within the flow of functional genomic research (Figure 2.2), bioinformatics is where targets from the initial large-scale screen that have been validated *post hoc* and tested

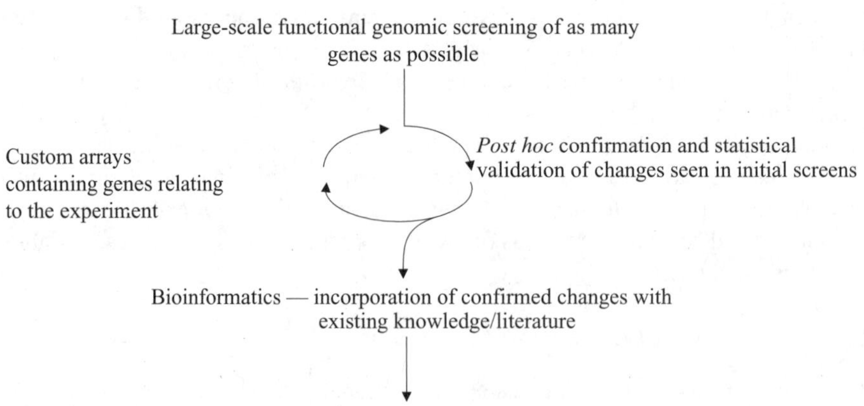

FIGURE 2.2 Functional genomic analysis is designed to gain a global perspective of gene expression in a particular experimental state. Functional genomic analysis begins with the screening of as many genes as possible to see what genes are expressed in the cells of interest in a particular condition, and what differences in gene expression may be of importance. To overcome the lack of statistical power and the large possibility of false positives with arrays, some form of *post hoc* testing is needed. Changes seen and confirmed in the hybridization array then need to be incorporated into the existing knowledge of the question at hand. Finally, to show direct causative links, interference or manipulation studies are needed.

on custom arrays begin to form a biological narrative. While the amount of data generated from functional genomic research is amazing, databases and clustering charts are not the ultimate goal of this research. Combining the existing knowledge of specific gene functions, the previous work on the subject, and the gene expression array data should result in a descriptive biological story. This may seem to be an obvious point, but in the excitement to use this new technology, the old rules of research should not be forgotten. To this end, new technologies and databases are currently being developed that will permit integration and mining of biological data for all genes, gene families, chromosome locations, and ESTs.

2.3.5 Dynamic Intervention/Target Validation

Traditionally, the gold standard for biological research has been to interfere with a biological phenomenon to show causative nature. Approaches used in this manner include gene knockout mice, antisense knockout approaches, specific protein inhibitors, and antagonists. Therefore, a key consideration is that once a gene has been illuminated by array analysis and its change confirmed by *post hoc* methods, a dynamic intervention should be conducted to confirm the direct involvement of the gene in the biology under study.

2.4 TECHNICAL CONSIDERATIONS IN ARRAY EXPERIMENTAL DESIGN

All successful science is based on sound experimental design. From a practical standpoint, this is especially true of hybridization array experiments because the time and resources that can be wasted on poorly designed functional genomic research are staggering. For both the beginning researcher and those already conducting experiments using hybridization arrays, it is worth examining the concerns of sample collection, sensitivity, *post hoc* confirmation, and data analysis (Figure 2.3).

2.4.1 Sample Collection

Sample collection is a basic element of experimental design for many molecular biological experiments, but it is worth reiterating. Specifically, given the expense of array analysis (in time, money, and energy), it is wise to invest considerable effort in determining that: (1) the key experiment is well conceived; and (2) the input samples are intact and appropriately prepared. Depending on the cells or tissue being examined, it is often unavoidable that a sample will contain multiple cell types. In complex samples, such as brain tissue, there is routinely a heterogeneous cell population. Therefore, observed changes may represent a change in one cell type or all cell types. Similarly, smaller changes occurring in only one type of cell may be hidden. Thus, researchers must be mindful of heterogeneous cell populations when drawing conclusions. Similarly, in comparing normal and cancer samples, there will be obvious differences in the proportion of the cell types (i.e., cancer cells will be overrepresented). Therefore, interpretations of differences in gene expression may be complicated by the sheer mass of one cell over another. A promising technological

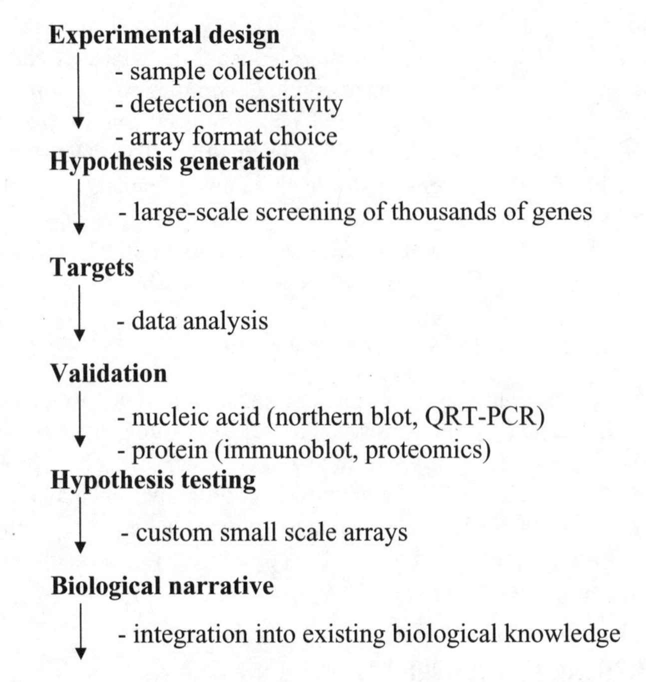

FIGURE 2.3 Technically, functional genomic analysis goes through three stages — hypothesis generation, target validation, and hypothesis testing — to arrive at the end-point of all functional genomic research: the biological narrative. Large-scale arrays (thousands of genes) are useful for initial screens of gene transcription. Changes seen on the hybridization array need to be validated by either nucleic acid or protein methodologies. To further investigate the question, custom or small-scale arrays can be constructed that contain the genes initially identified to be changed, as well as related genes. Ultimately, these gene expression changes can be incorporated into existing knowledge about the individual genes and the experimental question.

solution to this problem is laser capture microdissection, which allows very small and identified cellular populations to be dissected.[9] The amount of sample and RNA collected in this manner is so small, however, that either target or signal amplification steps must be used.[10,11]

The timing of tissue collection goes hand-in-hand with the nature of the collected tissue and therefore sample collection times will be important. For example, in an experiment in which cells undergo programmed cell death, the collection time point will determine if causative changes or end-point changes are to be observed. If a late time point is chosen, it becomes increasingly difficult to distinguish changes due to the general breakdown of cellular processes from those that have triggered the cell death.

An important issue in DNA array analysis is the use of individual samples or pooled RNA preparations from a number of samples. The pooling of equal amounts of RNA from all of the representatives of an experimental or control group (whether cells or animals) produces what can be termed an *expression mean*. The alterations in gene expression, illuminated by the resulting array analysis, reflect changes that are common to most/all animals or samples in a group. The outlier expression of one gene in a given animal/sample is therefore averaged toward unity. Of course on the converse side, unique responses that appear in a given animal and that might be quite relevant to the specific response of that animal are also lost. This aspect of experimental design, however, is intended to maximize the chances of a legitimate "hit" in the initial analysis. The result of this approach is that more of the target genes generated from the initial screen are statistically confirmed in *post hoc* testing. Finally, the cost of the array technology also necessitates consideration of pooled analysis, because it is often prohibitively expensive to perform experiments on individual animals/samples.

The most important component of a successful array experiment is the isolation and characterization of intact RNA. The common method for RNA isolation is the guanidinium thiocynate procedure.[12] Modifications of this protocol have been developed,[13] and the relative merits of this and other techniques have been reported.[14] RNA should always be subjected to denaturing gel electrophoresis to visually verify the integrity of the RNA by 28S and 18S ribosomal RNA bands and spectrophotometric measurements of RNA concentration have been reported to be sensitive to pH.[15] The same denaturing gel used to confirm integrity can also be used to visually verify the spectrophotometric quantification. Although this is such a basic aspect of all functional genomic analysis, it is worth reiterating the importance of careful sample preparation. RNA degradation is a serious technical problem and can lead to variable results. Although ribonuclease levels vary by organism and tissue, careful RNA isolation will enhance the subsequent output from the array.

2.4.2 Detection Sensitivity

There are two key detection issues when thinking about DNA hybridization arrays. The first is whether or not an mRNA can be detected (threshold sensitivity), and the second is whether or not changes in mRNA level will be large enough to be detected (fold-change sensitivity). These considerations will determine decisions about what platform to use, the use of poly (A^+) or total RNA, and detection methods (radioactivity or fluorescence).

2.4.2.1 Threshold Sensitivity

Detection sensitivity in array research takes two very distinct forms. The first, termed threshold sensitivity, is the ability to detect one RNA species out of a population and is a concern for rarely expressed messages, for small sample sizes, and is the traditional issue of sensitivity common to other techniques. Array analysis, when it does not involve signal amplification, is not the most sensitive method. This is in contrast to transcription-based aRNA amplification, or PCR-based differential display or quantitative RT-PCR. Therefore, levels of detection (the number of copies of a specific gene needed per unit of RNA to yield a signal) are not particularly sensitive. A number of approaches have been developed to increase signal output (RNA amplification, poly (A^+) RNA isolation, output signal amplification [sandwich detection methodologies]). However, every amplification procedure comes at the cost of variable amplification efficiencies and so extreme care must be taken in adopting these approaches. Unfortunately, there has been very little systematic comparison of array platforms and detection methods.[16] There is anecdotal evidence that membrane- and radioactivity-based macroarrays are more sensitive. However, there are unique concerns with the use of radioactivity and the macroarrays are generally perceived as less valuable because they screen fewer genes and generally do not provide widespread EST arrays for gene discovery.

2.4.2.2 Fold-Change Sensitivity

The second sensitivity parameter is *fold-change sensitivity*, or the ability of hybridization arrays to reliably determine a certain magnitude difference in expression. The claimed fold-change sensitivity of different platforms varies. Determination of this parameter is crucial to characterizing the technology and ensuring that researchers choose the technology most appropriate to their goals. For research involving systems that undergo large gene expression changes (e.g., yeast cell-cycle regulation, or organ developmental processes in which tenfold changes are expected), one can detect such changes with fluorescent protocols. Other research efforts, for example in neuroscience, where gene changes are less dramatic, may find radioactivity-based methods more applicable.

2.4.3 *Post hoc* Confirmation

One of the most common criticisms of hybridization arrays is that when hundreds or thousands of genes are examined at once, some apparent changes are the result of random chance. This is because a single array experiment, representing an *n* of one, lacks the sample size needed for statistical analysis. Indeed, at their core, most arrays essentially represent 1000 to 10,000 t-tests. As such, one is likely to find small magnitude changes (less than twofold) in signals that are not reflective of actual changes in mRNA levels. This is a statistical reality and highlights the requirement for *post hoc* confirmation of changes seen with arrays. So, how does one separate *bona fide* changes from type I statistical errors (false positives)? Tests on individual samples themselves are necessary to produce statistical significance. Such corroborating experiments can examine the gene changes at the level of mRNA (Northern blot, QRT-PCR), protein (immunoblot), or activity (enzymatic activity, DNA binding, or other measures). The protein and activity tests are recommended because they assess the gene of interest at a level closer to the function of the protein or actually address the function itself. Protein analysis is important because increased levels of transcription do not always translate into increased levels of protein.[3] In addition, protein assessment is achieved with fundamentally different experimental techniques and may not therefore be subject to the same sources of error as the array. Unfortunately, immunoblotting and activity assays would appear to return researchers to the single gene assay that hybridization arrays were intended to avoid. This is not true in practice, however, because large numbers of genes have already been screened by the array (see Figure 2.2). The optimal solution to ascribing relevance to the data is to develop techniques by which confidence intervals for individual genes can be generated from arrays and these results can be combined with proteomic techniques under development.[2] Alternatively, as costs are decreased, individual hybridization array experiments will be performed for each sample. As well, many researchers are exploring the use of small (in the number of genes) arrays that focus on a specific gene family or pathway.

Hybridization array technology has opened exciting new avenues of biomedical research. With this excitement, a sober view of experimental design is required. Truly groundbreaking research will require the same, if not greater, attention to experimental design than was required in the past. Because of the large effort and investment required for functional genomic research poorly conceived experiments can squander, it is worth considering these issues before undertaking major investments of time and resources.

2.4.4 Data Analysis

The creation, hybridization, and detection of microarrays can seem like a daunting task. It would appear that once an image of the array, with relative densities for each sample, has been generated, the experiment would nearly be finished. Unfortunately, this is not the case as scientists are now learning that the massive amounts of data generated by arrays pose a new challenge.[17–19] In this section, basic data analysis, computational models, and integration of data with existing biological knowledge will be examined.

2.4.4.1 Data Analysis Basics

The first steps in data analysis are background subtraction and normalization. The principles of both are similar to the techniques used with conventional nucleic acid or protein blotting. Background subtraction pulls the nonspecific background noise out of the signal detected for each spot and allows comparison of specific signals. For illustration, if the signal intensities for the control and experimental spots are 4 and 6, respectively, it would appear that the experimental value is 50% higher. However, if a background of 2 is subtracted from both signal intensities, the experimental value is actually 100% higher than control. A complication to background subtraction is that differences in background across the array can affect some spots more than others and therefore a local background from the area around each spot is often used.

Normalization is the process that accounts for the differences between separate arrays. All macroarray (membrane-based radioactively detected arrays) experiments and any other multiple array experiments may require the use of normalization for consistent comparisons. For example, when a pair of macroarrays representing control and treated samples show a difference in overall or total signal intensity, such differences can arise from unequal starting amounts of RNA or cDNAs, from different efficiencies of labeling reactions, or from differences in hybridization. Any of these factors can skew the results. Common methods of normalization include: a housekeeping gene(s), a gene thought to be invariant under experimental conditions; using the sum of all signal intensities; or a median of signal intensities. Housekeeping genes do in fact vary under some experimental conditions and are problematic for many experiments. All of these approaches have limitations and exogenous synthetic RNA standards have been used for normalization.[20]

2.4.4.2 Computational Methods

The sheer quantity of data generated by arrays exceeds the ability of manual human assessment. Advances in computational biology and bioinformatics are being used to effectively and exhaustively explore hybridization array results and create a biological story out of the databases generated by hybridization array data.[21,22]

In binary experiments where only one control and one experimental sample are being compared, the data analysis requires only a ratio of control to treated. For more involved experimental designs where there are two or more experimental conditions (typically, an experiment looking at multiple time points, doses, or groups), the computational requirements are much greater. The question is not one of a simple change under one condition, but becomes how does one gene (out of thousands) change over multiple conditions. With large experiments analyzing thousands of genes, the data increases dramatically and, as a result, it can be difficult to find patterns in the data. To this end, computational algorithms are used. These approaches seek to find groups of genes (clusters) that behave similarly across the experimental conditions. Clusters, and the genes within them, can subsequently be examined for commonalities in function or sequence to better understand how and why they behave similarly. A number of different methods — k-means, self-organizing maps, hierarchical clustering, and Bayesian statistics — are employed for clustering analysis.[23–25] Clustering analyses will be critical for the mining of public expression databases that are being generated.[26]

2.4.4.3 Integration with Other Biological Knowledge

In the excitement of using functional genomic technology, it is important to not forget what we already know about other biological measures. This is accomplished using the existing knowledge of genes and their functions and combining gene expression data with chemical, biochemical, and clinical measures. One example of combining gene expression data with other measures comes from the cancer field in the recent work by Alizadeh et al.[27] In this work, large B-cell lymphomas were put into subtypes by their gene expression profile, and these subtypes were found to have significantly different reactions to therapy.

2.5 CONCLUSION AND FUTURE DIRECTIONS

Undeniably, functional genomics is opening new avenues of research. The advances in technology that have made this possible are exciting in themselves and require a great deal of effort to perfect. In this climate, it is easy to succumb to technical showmanship and produce complex works that highlight the technology. While these are interesting works, the goal of most researchers is to increase biological knowledge for humanity. The fruits of functional genomic research will go to those who not only master the new technology, but also integrate these tools into well-designed experimental projects.

ACKNOWLEDGMENTS

This work was supported by NIH grants P50DA06643, P50AA11997, and R01DA13770 (to K.E.V.), and T32DA07246 (to W.M.F.).

REFERENCES

1. Lockhart, D.J. and Winzeler, E.A., Genomics, gene expression and DNA arrays, *Nature*, 405, 827, 2000.
2. Pandey, A. and Mann, M., Proteomics to study genes and genomes, *Nature*, 405, 837, 2000.
3. Anderson, L. and Seilhamer, J., A comparison of selected mRNA and protein abundances in human liver, *Electrophoresis*, 18, 533, 1997.
4. Mir, K.U., The hypothesis is there is no hypothesis, *Trends in Genetics*, 16, 63, 2000.
5. Freeman, W.M., Robertson, D.J., and Vrana, K.E., Fundamentals of DNA hybridization arrays for gene expression analysis, *BioTechniques*, 29, 1042, 2000.
6. Freeman, W.M., Walker, S.J., and Vrana, K.E., Quantitative RT-PCR: pitfalls and potential, *BioTechniques*, 26, 112, 1999.
7. Cheung, V.G., Morley, M., Aguilar, F., Massimi, A., Kucherlapati, R., and Childs, G., Making and reading microarrays, *Nat. Genetics*, 21, 15, 1999.
8. Schena, M., Ed., *DNA Microarrays: A Practical Approach* (Practical Approach Series), Oxford University Press, Oxford, 1999.
9. Luo, L. et al., Gene expression profiles of laser-captured adjacent neuronal subtypes, *Nat. Medicine*, 5, 117, 1999.
10. Van Gelder, R.N., von Zastrow, M.E., Yool, A., Dement, W.C., Barchas, J.D., and Eberwine, J.H., Amplified RNA synthesized from limited quantities of heterogeneous cDNA, *Proc. Natl. Acad. Sci. U.S.A.*, 87, 1663, 1990.
11. Wang, E., Miller, L.D., Ohnmacht, G.A., Liu, E.T., and Marincola, F.M., High-fidelity mRNA amplification for gene profiling, *Nat. Biotech.*, 18, 457, 2000.
12. Chomczynski, P. and Sacchi, N., Single-step method of RNA isolation by acid guanidinium thiocyanate-phenol-chloroform extraction, *Anal. Biochem.*, 162, 156, 1987.
13. Puissant, C. and Houdebine, L.M., An improvement of the single-step method of RNA isolation by acid guanidinium thiocyanate-phenol-chloroform extraction, *BioTechniques*, 8, 148, 1990.
14. Yamaguchi, M., Dieffenbach, C.W., Connolly, R., Cruess, D.F., Baur, W., and Sharefkin, J.B., Effect of different laboratory techniques for guanidinium-phenol-chloroform RNA extraction on A260/A280 and on accuracy of mRNA quantitation by reverse transcriptase-PCR, *PCR Methods Appl.*, 1, 286, 1992.
15. Wilfinger, W.W., Mackey, K., and Chomczynski, P., Effect of pH and ionic strength on the spectrophotometric assessment of nucleic acid purity, *BioTechniques*, 22, 474, 1997.

16. Baldwin, D., Crane, V., and Rice, D., A comparison of gel-based, nylon filter and microarray techniques to detect differential RNA expression in plants, *Curr. Opin. Plant Biol.*, 2, 96, 1999.
17. Bassett, D.E., Eisen, M.B., and Boguski, M.S., Gene expression informatics — it's all in your mine, *Nat. Genetics,* 21, 51, 1999.
18. Brent, R., Functional genomics: learning to think about gene expression data, *Curr. Biol.*, 9, R338, 1999.
19. Vingron, M., and Hoheisel, J., Computational aspects of expression data, *J. Mol. Med.*, 77, 3, 1999.
20. Eickhoff, B., Korn, B., Schick, M., Poustka, A., and van der Bosch, J., Normalization of array hybridization experiments in differential gene expression analysis, *Nucl. Acids Res.*, 27, e33, 1999.
21. Claverie, J.M., Computational methods for the identification of differential and coordinated gene expression, *Human Mol. Gen.*, 8, 1821, 1999.
22. Zhang, M.Q., Large-scale gene expression data analysis: a new challenge to computational biologists, *Genome Res.*, 9, 681,1999.
23. Ben-Dor, A., Shamir, R., and Yakhini, Z., Clustering gene expression patterns, *J. Computational Bio.*, 6, 281, 1999.
24. Hilsenbeck, S.G., Friedrichs, W.E., Schiff, R., O'Connell, P., Hansen, R.K., Osborne, C.K., and Fuqua, S.A.W., Statistical analysis of array expression data as applied to the problem of tamoxifen resistance, *J. Natl. Cancer Inst.*, 91, 453, 1999.
25. Tamayo, P., Slonim, D., Mesirov, J., Zhu, Q., Kitareewan, S., Dmitrovsky, E., Lander, E.S., and Golub,T.R., Interpreting patterns of gene expression with self-organizing maps: methods and application to hematopoietic differentiation, *Proc. Natl. Acad. Sci. U.S.A.*, 96, 2907, 1999.
26. Claverie, J.M., Do we need a huge new centre to annotate the human genome? *Nature*, 403, 12, 2000.
27. Alizadeh, A.A. et al., Distinct types of diffuse large B-cell lymphoma identified by gene expression profiling, *Nature*, 403, 503, 2000.

3 Oligonucleotide Array Technologies for Gene Expression Profiling

David Dorris, Ramesh Ramakrishnan, Tim Sendera, Scott Magnuson, and Abhijit Mazumder

CONTENTS

3.1 INTRODUCTION

The evolution of Southern blots into filter-based screening and, with the incorporation of high-speed robotic printing, miniaturization, and fluorescence detection technologies, into the microarrays of today has created a new era in systems biology and therapeutics. Two avenues are available for microarrays, employing either amplified cDNAs (generally 0.5 to 2 kb in length) or oligonucleotides on the array. This review focuses primarily on oligonucleotide array technologies, performance, and applications.

3.2 ADVANTAGES AND DISADVANTAGES OF OLIGONUCLEOTIDE ARRAYS

When gene sequence information is available, oligonucleotides can be designed and synthesized to hybridize specifically to each gene in the sample. This approach

0-8493-2285-5/02/$0.00+$1.50

obviates the need for management (tracking and handling) of large clone libraries because it is guided primarily by sequence data. Implicit in this statement is the fact that PCR amplification (and the associated labor and costs) and sequence verification are no longer necessary. Furthermore, the ease of *in silico* design and the specificity of oligonucleotides enable representation (on the array) and discrimination of rarely used splicing patterns (which would be hard to find as cloned cDNAs) and allow one to distinguish between closely related (and possibly differentially regulated) members of gene families. An example of this level of specificity using Motorola Codelink™ Expression chips is shown in Section 3.5. We designed probes to the alcohol dehydrogenase genes 1 and 2 whose expression levels were previously shown to be indistinguishable by cDNA arrays[1] because of their high level of sequence homology (88%).

Oligonucleotide arrays are particularly well suited to analyze the expression profiles of organisms with completely sequenced genomes[2,3] because all predicted genes and exons can be analyzed. Two elegant examples of this approach were recently demonstrated for the *E. coli*[4] and human genomes.[5] The former study generated an array containing, on average, one 25mer probe per 30-bp region over the entire *E. coli* genome. This high-resolution array relied on genomic sequence, rather than sequence derived from ESTs (expressed sequence tags), to generate oligonucleotide probes and analyze operon structure and the corresponding mRNAs. The latter study generated an array consisting of 50 to 60mer oligonucleotide probes derived from all predicted exons to validate predicted exons, group exons into genes as determined by co-regulated expression, and define full-length mRNA transcripts. This study also generated a high-resolution tiling array with overlapping probes to various genomic regions on chromosome 22, which could reveal exons not identified *in silico* and provide information about exon structure and splicing.

However, disadvantages of oligonucleotide arrays also exist. For example, oligonucleotides consisting of 20 to 30 bases in length exhibit reduced sensitivity when using fluorescently labeled, first-strand cDNA generated from the RNA sample. This limitation can be overcome using various amplification schemes[6] or oligonucleotides of 50 bases in length.[7] Further disadvantages depend on methods of oligo array fabrication (see Section 3.2 for detailed array production methods). For example, for *in situ* synthesis, step-wise yield and methods to QC (quality control) the final product on the array can limit purity and impact specificity and sensitivity in an assay. For arrays where the oligonucleotide is synthesized off-line and deposited on the array, the cost of oligonucleotide synthesis, the (oligonucleotide length-based) need for covalent attachment schemes to prevent washing the oligonucleotides off the array, and high-throughput tracking and QC of oligonucleotides prior to deposition can impact array production. However, several innovative solutions in chemistry and systems engineering have been proposed to address these obstacles.[8,9]

3.3 ARRAY FABRICATION TECHNOLOGY

Oligonucleotides can be synthesized *in situ* or prefabricated and then printed. Synthesis of oligonucleotides by light-directed, combinatorial solid-phase chemistry[10] or other *in situ* methods[11,12] offers the advantage of having the oligonucleotide synthesized on the support that will be used in the hybridization, obviating the need to hydrolyze the oligonucleotide from its synthetic support and reattach it to the microarray. There are disadvantages to this approach as well. First, it does not allow an independent confirmation of the fidelity of synthesis. Second, due to the lower yields of many of these *in situ* synthetic protocols, oligonucleotides synthesized have generally not been longer than 25 bases. Third, this approach does not allow purification of the oligonucleotide prior to attachment to the microarray. Fourth, because the oligonucleotides are attached to the support at their 3′ ends, they cannot be used in polymerase-mediated extension reactions. However, a change to the synthetic scheme has been proposed to address the latter two issues.[13]

A powerful version of *in situ* synthesis employs photolithographic microarray design. Photolithography enables the large-scale production of extremely high-density arrays wherein the sequence of the oligonucleotides synthesized at each distinct feature is independently directed. However, the success of this method is dictated by mechanical (accuracy in alignment of photomasks) and chemical (efficiency of photoprotecting group removal and phosphoramidite coupling) factors. Recent technological advances in surface patterning,[14] electrochemistry, optics,[15] and synthetic chemistry[16] have enabled new methods of *in situ* synthesis with fewer limitations with respect to oligonucleotide length, cost, equipment, array redesign (flexibility), and time required for array fabrication (see Table 3.1 for an index of array companies and fabrication methods). Some of these *in situ* synthesis methods are particularly amenable to rapid probe prototyping prior to final array design and thus are powerful research tools as well.

Covalent attachment of prefabricated oligodeoxyribonucleotides circumvents some of the constraints imposed by earlier *in situ* synthesis methods and allows new elements to be added without redesigning the entire microarray. The primary concern with postsynthetic attachment is whether a robust, specific, irreversible, and reproducible attachment chemistry can be created to yield high sensitivity and reproducibility in the subsequent assays. Our laboratory has demonstrated fabrication of arrays by photochemical as well as chemical attachment.[17] Incorporation of specific functional moieties at the 5′ end of oligonucleotides can serve as a pseudo-purification step if nonspecific adsorption of the oligonucleotide is eliminated because only full-length oligonucleotides will receive the attachment group and will be the only ones that attach to the matrix efficiently.

Noncovalent retention of oligonucleotides can be exploited when longer oligos (e.g., 60 to or 70mer) exhibit similar chemical characteristics as cDNAs (i.e., they generate sufficient electrostatic interactions and contain sufficient numbers of pyrimidines to enable efficient crosslinking). Oligonucleotides retained on a glass surface in this manner may not exhibit the same degrees of conformational flexibility or accessibility as do those retained via end attachment. However, noncovalent retention does obviate the need for special chemistries and, because longer oligonucleotides are used, may eliminate the need for target amplification schemes. Alternatively, oligonucleotides can be anchored by high-affinity interactions such as biotin-streptavidin. This scheme enables end attachment of oligonucleotides without special chemistries, but conditions under which the biological interaction remains intact may impose constraints upon subsequent hybridization or array processing.

TABLE 3.1
Array Fabrication Methods and Companies

Company	URL	Notes
		***In Situ* Synthesis**
Agilent/Rosetta	www.agilent.com	Inkjet printing
Affymetrix	www.affymetrix.com	Solid-phase chemical synthesis with photolithographic fabrication
FeBit	www.febit.com	Light-directed synthesis in a parallel sample processor
Nimblegen	www.nimblegen.com	Virtual masks relayed to a digital micromirror array
Combimatrix	www.combimatrix.com	Porous reaction layer coating a semiconductor surface coupled with virtual flask synthesis
Protogene	www.protogene.com	Photoresist-mediated surface patterning prior to oligo synthesis
		Covalent Attachment
Illumina	www.illumina.com	Fiber-optic bundles containing self-assembled microsphere arrays
Motorola	www.motorola.com	Piezoelectric dispensing of oligos onto a porous 3-D surface
Mergen	www.mergen-ltd.com	Pre-spotted oligo arrays
		Noncovalent Deposition
Genometrix	www.genometrix.com	Low-density genes printed in triplicate
Nanogen	www.nanogen.com	Electronic addressing of biotinylated molecules to a streptavidin-agarose permeation layer
Operon	www.operon.com	70mer arrays

3.4 PROBE DESIGN CONSIDERATIONS

Oligonucleotide probes are generally designed to the 3′ end of an RNA transcript to eliminate any uncertainty and possible complications of transcript degradation.[18] The more unique sequence identity of the 3′ untranslated region of a transcript from a gene family may allow easier discrimination relative to the similar nature of coding regions of gene family members. In addition, priming and amplification schemes (random hexamer vs. oligo-dT) can also impact which regions of the transcript will be represented in the cDNA or cRNA sample, guiding probe design as well. Finally, technological limitations of cRNA generation of long transcripts (e.g., reverse transcriptase and RNA polymerase processivity) may be better addressed by designing oligonucleotide probes to the 3′ regions of RNAs. Although a set of heuristics has been proposed for probe design,[19] there is a general lack of data regarding determinants of effective oligonucleotide hybridization to cDNA or cRNA targets. This scenario underscores the importance of rapid probe prototyping and/or the use of multiple probes per transcript in expression profiling arrays. Basic studies on heteroduplex formation as it pertains to microarrays are now underway,[20] and analogies to antisense oligonucleotides (whose efficacy depends on hybridization and transcript cleavage) may yield further insights and heuristics.[21]

The general method used for oligonucleotide probe design in our laboratory is detailed in Figure 3.1 and in the following paragraphs. Specific oligonucleotide probes are generated from EST and genomic sequence databases and used for identification and quantitation of specific gene products on microarray platforms. In the case of Motorola microarray products, the oligonucleotide capture probes are synthesized by standard phosphoramidite chemistry, validated for purity and sequence accuracy, and then deposited on to a polymer-coated glass surface using piezoelectric noncontact printing. The suitability of an expression probe is governed by a number of factors, including biophysical characteristics of the oligonucleotide probes and of the intended targets.

Initially, all available ESTs and mRNA sequences are clustered and aligned using standard bioinformatics tools to generate a set of consensus sequences representing a single, unique, high-quality sequence for each potential gene target. These assembly methods are designed to distinguish between individual genes with high sequence homology, as well as identifying single genes that contain several alternatively spliced transcripts. The consensus sequences are then examined and regions of single nucleotide repeats, low complexity, genomic sequence, and possible polymorphisms are masked and avoided during the probe design process.

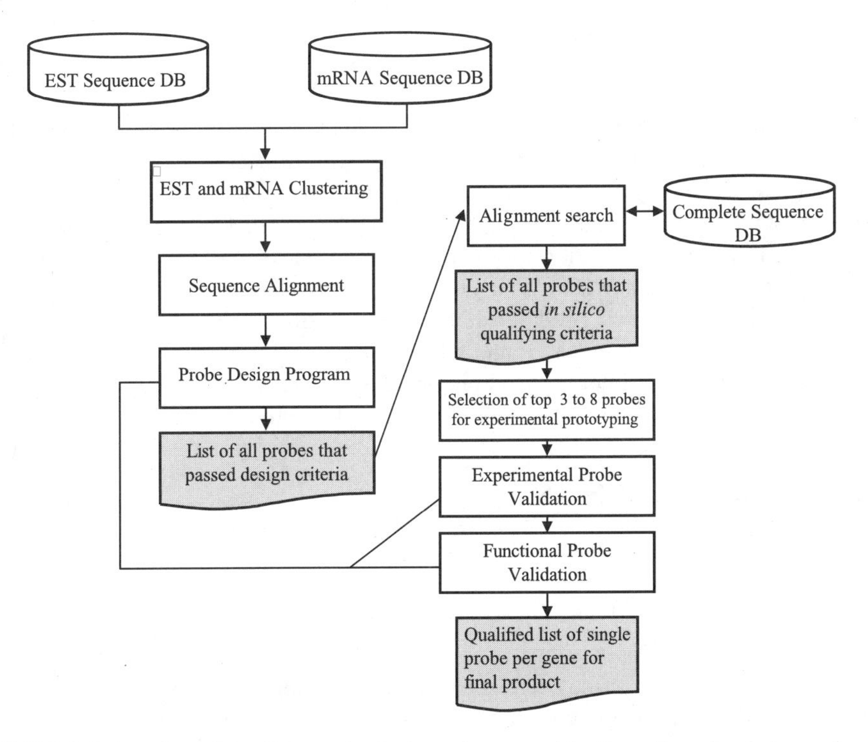

FIGURE 3.1 Overview of the probe design scheme used by the authors for design, selection, and validation of oligonucleotide probes.

The high-quality consensus sequences generated for each gene are then scanned for all possible 30mer strings that possess several predefined biophysical characteristics. Some of the parameters used to identify a good-quality probe for microarray analysis include distance from the 3′-end, consistent Tm (melting temperature for the hybrid), GC content, and free energy of probe-target hybrid. Potential probes are examined and omitted if they demonstrate a high probability for secondary structures or probe-probe duplex formation; both situations would interfere with probe-to-target interactions. Each of these parameters will vary, depending on the platform substrate as well as the desired hybridization conditions. Finally, if target fragmentation steps are not employed, target secondary structure formation should be used to identify target regions of low instances of intermolecular hybridization.

Once several candidate probes are generated for each gene of interest, probe sequences are compared against the most complete gene databases with sequence alignment tools such as BLAST or FASTA as an *in silico* verification of probe selectivity and specificity for the intended target. Following this filtering, up to six candidate probes with the predicted specificity are selected, synthesized, and tested for experimental specificity and selectivity against a panel of selected tissues. In general, the requested top three rated probes based on the *in silico* design parameters are synthesized and deposited onto prototyping chips termed "*screen design chips*." Each screen design chip contains three probes per gene of interest and is processed by the standard protocol with one exception. In the screen procedure, two different target concentrations are used per target tissue tested. The single best oligonucleotide probe based on empirical evidence is then retained to represent each particular gene for the final microarray product.

This paradigm for probe design and selection is applicable for oligonucleotide probes of any length. Flexibility in probe design with respect to length is important because k_2, the second-order rate constant, is proportional to the square root of the length of the shortest strand participating in duplex formation.[22] A more detailed discussion of the hybridization reaction is given in Section 3.4. A recent study by the Rosetta team has demonstrated the utility of 60mers fabricated *in situ* for expression analyses and shown good sensitivity under various hybridization conditions (although good specificity was found when at least 18 mismatches in the 60 base region were present).[23] With respect to probe design, that study also reported higher cross-hybridization with oligonucleotides enriched with deoxycytidine.

3.5 HYBRIDIZATION AND DETECTION

The hybridization rate constant, k_2, is described by

$$k_2 = [k'_N (L_S)^{1/2}]/N \tag{3.1}$$

where L_S is the length of the shortest strand participating in duplex formation, N is the complexity (total number of base pairs present in nonrepeating sequences), and k'_N is the nucleation rate constant. Although this equation has not been rigorously verified for oligonucleotides, estimates based on this equation are very accurate. The nucleation rate constant is affected by temperature, ionic strength, and viscosity and can be 20 to 50% lower for RNA-DNA hybridization vs. DNA-DNA hybridization, depending on the amount of RNA secondary structure. The maximum hybridization rate occurs when the difference between the hybridization temperature and the Tm is 25°C.[22] However, when the difference is this high, implying low hybridization temperatures, the dissociation rates are also decreased considerably. Therefore, other measures (see below) may be required to ensure adequate specificity in the hybridization.

Because of the dependence of the k'_N on ionic strength, hybridization reactions are often performed at high salt concentrations. To maintain the stringency at this high salt concentration, denaturants such as formamide are generally added or high temperatures, which may introduce evaporation problems, are needed. The k'_N, however, can be reduced if the denaturing solvent has high viscosity. For example, k'_N decreases 1% for every 1% addition of formamide.

Labeling of cDNA and cRNA can be done by several methods, dictating direct or indirect detection methods. Cyanine-labeled dNTPs or NTPs can be incorporated by reverse transcriptases or RNA polymerases.[24] Alternatively, biotin or other protein/antibody binding moieties (dinitrophenol, digoxigenin, etc.) can be incorporated, followed by detection with a streptavidin- or antibody-based method. There are several advantages of biotin incorporation. First, biotin-labeled nucleotides are efficient substrates for many DNA and RNA polymerases. Second, cDNAs or cRNAs containing biotinylated nucleotides have denaturation, reassociation, and hybridization parameters similar to those of unlabeled counterparts.[25] Third, the effect on yield of cDNA and cRNA can be less than that seen when cyanine dyes are incorporated into nucleic acids (D. Dorris, R. Ramakrishnan, and A. Mazumder, unpublished data). A third method of labeling is enzymatic incorporation of allylamine-derivatized dNTPs or NTPs, followed by derivatization with an amine-reactive derivative of a fluorophore or biotin.

The ability to label samples with multiple fluorophores introduces the question of one color vs. two-color hybridizations. The two-color approach offers several advantages. Using this approach, hybridization of two samples is performed on the same slide, eliminating the possibility that different spot morphologies, probe amounts, or inconsistencies in the hybridization could alter the ratio. Second, the PMT (photomultiplier tube) voltages can be adjusted in different channels to equalize intensity values on each slide. Third, CVs (coefficients of variation) in ratios are typically lower than CVs of raw hybridization signals.[23] However, the two-color approach also has disadvantages. For example, different fluorescently labeled nucleotides may be incorporated with different frequencies, altering the ratio due to an enzymatic parameter rather than a transcript abundance. Second, multiple experiment comparisons are not possible without replicating the reference sample (which, in some cases, may be difficult to obtain). Third, spectral overlap between dyes can complicate instrumentation or algorithms used in analysis. Fourth, executing signal amplification schemes in two colors is more complex than in single-color because multiple haptens are required.

Specificity during and after the hybridization reaction can be efficiently monitored through the use of negative controls (probes corresponding to bacterial or plant genes that do not cross-hybridize to the complex message applied to the array). We have developed a set of 84 negative control probes that can be used in low redundancy for the above-specified application as well as in high redundancy to verify uniformity in the hybridization reaction across the slide. Figure 3.2 shows an example of the negative controls and the calculation of a threshold based on the mean negative signal plus three standard deviations.

3.6 DATA QUALITY AND VALIDATION

Several metrics are useful for monitoring the performance of oligonucleotide arrays for expression profiling. For example, specificity can be measured by introducing mismatches into the oligonucleotide attached to the array and determining the number of mismatches that result in a loss of signal. Alternatively, specificity can be measured by the ability to discern expression levels of members of gene families. Figure 3.3 shows an example of both methods.

Sensitivity can be monitored by spiking transcripts into the complex message at mass ratios corresponding to those expected for low expressors. A mass ratio of 1 in 300,000 is generally used to represent a transcript having an abundance of one copy per cell.[26] Measuring the signal-to-negative control threshold ratio generally provides an accurate indication of the sensitivity of the platform.

Variability in array data can be assessed in a number of ways. CVs, correlation coefficients, and percent of data points within twofold when two replicates are compared (minimal detectable fold change) are methods currently employed in our

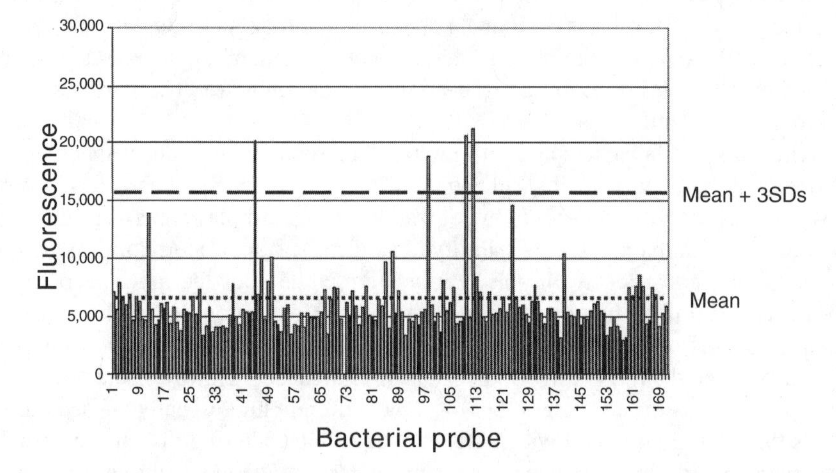

FIGURE 3.2 The use of approximately 170 negative controls to establish a threshold value. The probes to the orthogonal genes were designed in the same manner as the other probes on the array. These probes, however, should not hybridize to transcripts present in the complex mixture. The fluorescence intensity is shown on the *y*-axis. The mean and threshold (mean plus three standard deviations) are depicted by the dotted and dashed lines, respectively. Note that four probes have a signal that is above the threshold; these may represent cross-hybridization artifacts or true hybridization to rare transcripts (note the low signals) not present in GenBank.

laboratory. When CV is plotted as a function of intensity, the CV is found to decrease at higher intensities (D. Dorris, R. Ramakrishnan, and A. Mazumder, unpublished data). A trendline composed of a 50- to 100-probe moving average can be used to depict the general trend quite well. In many fluorescence-based measurements, fluorescent particles may alter the mean of several replicates, generating a high CV for that particular data point. We have found that plotting the CV as a function of the median intensity (more specifically, the median-normalized median intensity) serves to minimize the contribution of such anamolies (D. Dorris, R. Ramakrishnan, and A. Mazumder, unpublished data). The problem with using CV as a quality metric is that at least three replicates must be performed, thereby adding to the cost of the experiment. Correlation coefficients and the minimal detectable fold-change can be employed to examine a single or all possible pairwise array combinations. This metric, however, can also change at different intensity levels. For example, a fold change of 1.3 might be meaningful at high expression levels, whereas a fold-change of 1.8 might be considered within the noise at intensities near the negative control threshold. Others in the field have used *t*-tests (e.g., in the GeneSpring software from Silicon Genetics) to determine statistical significance of fold-changes. A recent study that proposed a method using normal distributions and posterior probabilities underscored the importance of replication in microarray studies.[27]

Validation of array data can be done in several ways. We have compared differential expression ratios generated by our platform to those generated by TaqMan (T. Sendera and S. Magnuson, unpublished data) and to those documented in the literature (e.g., from cDNA arrays). Finally, we have also compared expression ratios

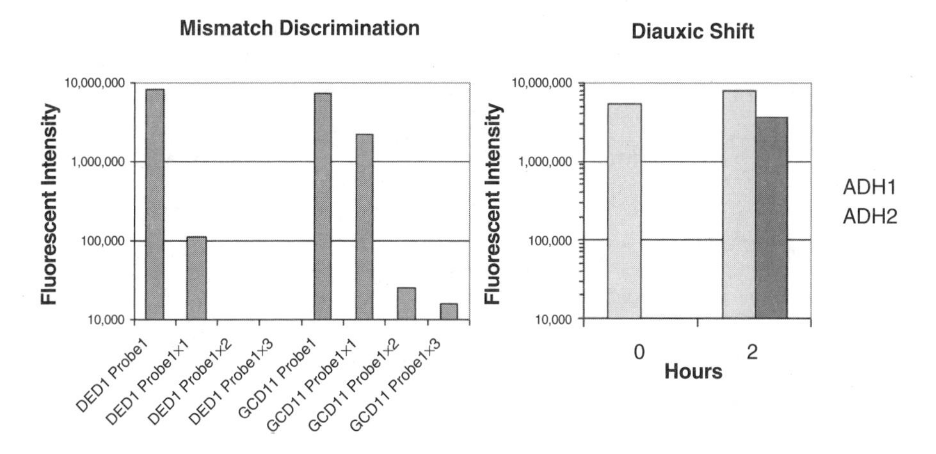

FIGURE 3.3 Specificity of hybridization of oligonucleotides. The left panel shows the decrease in fluorescence as centrally located mismatches are introduced into the oligonucleotide probe. The presence of one mismatch can have small (e.g., GCD11) or large (e.g., DED1) effects on the hybridization signal. However, three mismatches have been demonstrated to reduce hybridization signals to threshold. The right panel shows the differential regulation of the yeast ADH1 and two genes under conditions of diauxic shift. These genes are 88% identical at the nucleotide level. Their discernible expression changes are consistent with the ability to detect up to 90% specificity, as shown by the mismatches in the left panel.

to numbers of ESTs (expressed sequence tags) as a qualitative measure that transcript changes are being reported as expected (T. Sendera, S. Magnuson, and A. Mazumder, unpublished data). We have used the budding yeast *Saccharomyces cerevisiae* as a model eukaryotic organism as one method to validate our array data. Numerous transcriptional profiling studies have been performed on yeast, many of which are available in public databases.[1] Budding yeasts are also amenable to standard molecular techniques. Our yeast array contained a small number of genes at various expression levels whose abundance should or should not change under defined conditions. Figure 3.4 is a plot of transcript levels from cells grown under normal conditions compared to cells grown under heat shock conditions or to cells grown with an alternative carbon source. As expected, the heat shock gene *HSP12* is activated after the heat shock and the galactose and heat shock genes were induced after cells were grown in galactose. Interestingly, the magnitude of the fold-change in expression for *HSP12* was observed to be much greater than that reported for cDNA arrays.[1] We and others[23] have observed this phenomenom, wherein oligonucleotide arrays demonstrate a greater dynamic range in the reported differential expression ratios, particularly in the case of members of gene families. One possibility for this difference is the lack of cross-hybridization with oligonucleotide arrays. If only one member of the gene family is significantly differentially regulated under certain conditions, the ratio reported by oligonucleotides designed toward that particular gene may more accurately reflect the change in expression, whereas the cDNA element may reflect a weighted average of changes in differential expression for that gene and others to which it cross-hybridizes (which may or may not demonstrate changes in expression). Alternatively, a recent study published by the Incyte group[28] has demonstrated that the amount of PCR product arrayed can affect the differential expression ratios. This study showed that concentrations of less than 100ng/l result in a compression of the observed differential expression ratio. These

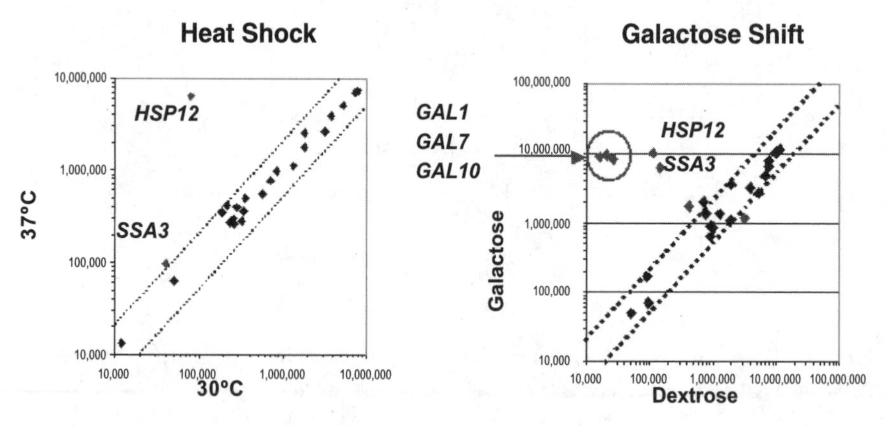

FIGURE 3.4 Differential expression changes in yeast induced by heat shock (left) or by galactose shift (right). Induction of the *HSP12* gene is consistent with data reported by cDNA arrays.[1] Induction of the GAL genes was as expected following shift to galactose. In addition, for this particular strain of yeast, the induction of *HSP12* and *SSA3* also occurs in response to this shift.

results, if generally applicable, may dictate normalization of PCR amplicon concentrations across amplicon batches in order to eliminate variance in the expression ratios due to concentration differences if ratios are taken across batches of arrays made from different PCR amplicon preparations. Further studies are required to determine the mechanism(s) resulting in such compression.

3.7 FUTURE WORK AND SUMMARY

With respect to applications, transcript profiling for target discovery, mechanism of action, and lead optimization studies will continue to be important in the drug discovery process. New applications, such as single-cell profiling, exon mapping,[5] and use of a subset of genes as candidate molecular markers in tumor classification,[29,30] diagnostic or prognostic applications,[31] or progression to metastatic phenotypes[32] will also impact the evolution of microarrays.

These more recent applications have demonstrated an important utility of microarray expression data as analytical, rather than purely descriptive, tools and have demonstrated the ability to use primary patient material. These changes have opened the door to new possibilities. For example, the use of expression profiling (and clinical data) in defining prognostic groups may allow earlier intervention in various treatment regimens for different patient populations. Furthermore, expression profiling could be used to complement genotyping technologies that are currently being considered for patient stratification in clinical trials.[33]

Increases in sample throughput, decreases in variability and labor, and decreases in the sample amounts required will drive new innovations in microarray technology from a performance standpoint. Increases in the extent of automation throughout sample preparation, hybridization, detection, and scanning, as well as improved manufacturing technologies, should help to address the first two concerns. Alternate target amplification or detection strategies may be required to further reduce RNA amounts.

A recently described fluorescence-based signal amplification, RCA (rolling circle amplification), has been developed by Molecular Staging Inc. (www.molecularstaging.com). This technology is driven by DNA polymerase and can replicate circular oligonucleotide probes with either linear or geometric kinetics under isothermal conditions. In the presence of two suitably designed primers, an exponential amplification occurs via DNA strand displacement and hyperbranching to generate 10^{12} or more copies of each circle in 1 hour. If a single primer is used, RCA generates a linear chain of thousands of tandemly linked DNA copies of a target covalently linked to that target in a few minutes. This linear RCA approach should permit accurate quantitation of a target. DNA generated by either the linear or exponential methods can be labeled with fluorescent DNP-oligonucleotide tags that hybridize to multiple sites in the product DNA sequences. The "decorated" DNA can then be condensed into a small point source by crosslinking with a multivalent anti-DNP IgM.

An example of a promising, nonfluorescence-based detection strategy that does not require target labeling is the electrical detection technology being pioneered by the Clinical Microsensors Division (www.microsensor.com) of Motorola Life

Sciences. This detection technology relies on bioelectronics (i.e., forming electronic circuit elements using assemblies of organic molecules on surfaces). The technology starts with a circuit board material that contains several electronically active sites, or electrodes. Oligonucleotide capture probes are attached to the gold-coated, electrically insulated electrodes on each biochip. The system also contains signaling probes with proprietary electronic labels attached to them. These signaling probes also bind to the target. Binding of the target sequence to both the capture probe and the signaling probe connects the electronic labels to the surface. In this manner, the complementary binding properties of DNA and RNA are used to assemble an electronic circuit element, generating a detectable electronic signal when the target DNA or RNA is present. When a slight voltage is applied to the sample following hybridization, the labels release electrons, producing a characteristic signal that can be detected through the electrode. Proprietary signal-processing technology is then used to quantify the amount of target present.

Other work that could reduce sample amounts include the use of modified nucleotides in either the probe or target molecules. These modified nucleotides would serve to increase the affinity of the probes for the targets. For example, by adopting a preorganized A-helical structure or enabling better hydration of the duplex formed with RNA, 2′-O-methyl- and 2′-fluororibonucleotides and N3′-P5′-phosphoramidate linkages are energetically favored (less negative entropy changes and more enthalpic stabilization) to bind cRNA targets with higher affinity than their unmodified 2′-deoxyribo counterparts.[34,35] Thus, oligonucleotides with torsion angles similar to those of A-form helices would be advantageous. Analogously, 2-aminoadenine (2,6-diaminopurine), 5-bromocytosine, and 5-methylcytosine have been shown to bind DNA targets with higher affinity than their unmodified 2′-deoxyribo counterparts. Because the phosphoramidite precursors of many modified nucleotides are now readily available, this approach is amenable to *in situ* synthesis and covalent attachment schemes. Alternatively, the modifications can be incorporated into the cDNA or cRNA targets. For example, 2,6-diaminopurine triphosphate can be incorporated by various polymerases. Enzymatic incorporation (binding affinities and polymerization rates) currently imposes a greater challenge than chemical incorporation, making incorporation into the oligonucleotide probe the method of choice.

In summary, oligonucleotide array platforms offer both advantages and disadvantages compared to cDNA arrays. From a practical perspective, the cost, performance, and accessibility of oligonucleotide arrays have limited their widespread use in the research market. From a geneticist's perspective, generation and amplification of cDNA libraries and clones are more familiar methodologies than oligonucleotide synthesis and attachment schemes. Motorola, among other companies, is seeking to address issues with oligonucleotide arrays and enable their widespread use. Once this occurs, specific applications requiring the ability to discern differential regulation of gene family members or alternative transcripts favor the use of oligonucleotide arrays. Furthermore, due to the lack of cross-hybridization of oligonucleotide probes on the array to transcripts present in the complex mixture, and perhaps due to other mechanisms, oligonucleotide arrays may offer a larger dynamic range in the fold-change of differential expression compared to cDNA arrays. Although this technology is still maturing and informatic methods and databases are still being developed to report, analyze, and standardize data generated, microarrays will empower researchers to discover new pharmaceutically relevant targets and elucidate functions and pathways of novel proteins.

REFERENCES

1. DeRisi, J.L., Iyer, V.R., and Brown, P.O., Exploring the metabolic and genetic control of gene expression on a genomic scale, *Science*, 278, 680–686, 1997.
2. Wodicka, L., Dong, H., Mittmann, M., Ho, M.-H., and Lockhart, D.J., Genome-wide expression monitoring in Saccharomyces cerevisiae, *Nat. Biotechnol.*, 15, 1359–1367, 1997.
3. Hill, A.A., Hunter, C.P., Tsung, B.T., Tucker-Kellogg, G., and Brown, E.L., Genomic analysis of gene expression in *C. elegans*, *Science*, 290, 809–812, 2000.
4. Selinger, D.W., Cheung, K.J., Mei, R., Johansson, E.M., Richmond, C.S., Blattner, F.R., Lockhart, D.J., and Church, G.M., RNA expression analysis using a 30 base pair resolution *Escherichia coli* genome array, *Nat. Biotechnol.*, 18, 1262–1268, 2000.
5. Shoemaker, D.D., Schadt, E.E. et al., Experimental annotation of the human genome using microarray technology, *Nature*, 409, 922–927, 2001.
6. Baugh, L.R., Hill, A.A., Brown, E.L., and Hunter, C.P., Quantitative analysis of mRNA amplification by *in vitro* transcription, *Nucl. Acids Res.*, 29, e29, 2001.
7. Kane, M.D., Jatkoe, T.A., Stumpf, C.R., Lu, J., Thomas, J.D., and Madore, S.J., Assessment of the sensitivity and specificity of oligonucleotide (50mer) microarrays, *Nucl. Acids Res.*, 28, 4552-4557, 2000.
8. Graves, D.J., Powerful tools for genetic analysis come of age, *Trends Biotechnol.*, 17, 127–134, 1999.
9. Marshall, W.S. and Boymel, J.L., Oligonucleotide synthesis as a tool in drug discovery research, *Drug Discovery Today*, 2, 34–42, 1999.
10. Fodor, S.P.A., Read, J.L., Pirrung, M.C., Stryer, L., Lu, A.T., and Solas, D., Light-directed, spatially addressable parallel chemical synthesis, *Science*, 251, 767–773, 1991.
11. Maskos, U. and Southern, E.M., Oligonucleotide hybridizations on glass supports: a novel linker for oligonucleotide synthesis and hybridization properties of oligonucleotides synthesized *in situ*, *Nucl. Acids Res.*, 20, 1679–1684, 1992.
12. Nanthakumar, A., Pon, R.T., Mazumder, A., Yu, S., and Watson, A., Solid-phase oligonucleotide synthesis and flow cytometric analysis with microspheres encoded with covalently attached fluorophores, *Bioconj. Chem.*, 11, 282–288, 2000.
13. Kwiatkowski, M., Fredriksson, S., Isaksson, A., Nilsson, M., and Landegren, U., Inversion of *in situ* synthesized oligonucleotides: improved reagents for hybridization and primer extension in DNA microarrays, *Nucl. Acids Res.*, 27, 4710–4714, 1999.
14. Chrisey, L.A., O'Ferrall, C.E., Spargo, B.J., Dulcey, C.S., and Calvert, J.M., Fabrication of patterned DNA surfaces, *Nucl. Acids Res.*, 24, 3040–3047, 1996.
15. Singh-Gasson, S., Green, R.D., Yue, Y., Nelson, C., Blattner, F., Sussman, M.R., and Cerrina, F., Maskless fabrication of light-directed oligonucleotide microarrays using a digital micromirror array, *Nat. Biotechnol.*, 17, 974–978, 1999.
16. Beier, M. and Hoheisel, J.D., Production by quantitative photolithographic synthesis of individually quality checked DNA microarrays, *Nucl. Acids Res.*, 28, e11, 2000.

17. Elghanian, R., Xu, Y., McGowen, J., Siethoff, M., Liu, C.G., Winick, J., Fuller, N., Ramakrishnan, R., Beuhler, A., Johnson, T., Mazumder, A., and Brush, C., The use and evaluation of 2+2 photoaddition in immobilization of oligonucleotides on a three dimensional hydrogel matrix, *Nucleosides and Nucleotides*, in press.
18. Lipshutz, R.J., Fodor, S.P.A., Gingeras, T.R., and Lockhart, D.J., High density synthetic oligonucleotide arrays, *Nat. Genetics* (*Supplement*), 21, 20–24, 1999.
19. Lockhart, D.J., Dong, H., Byrne, M.C., Follettie, M.T., Gallo, M.V., Chee, M.S., Mittmann, M., Wang, C., Kobayashi, M., Horton, H., and Brown, E.L., Expression monitoring by hybridization to high-density oligonucleotide arrays, *Nat. Biotechnol.*, 14, 1675–1680, 1996.
20. Mir, K.U. and Southern, E.M., Determining the influence of structure on hybridization using oligonucleotide arrays, *Nat. Biotechnol.*, 17, 788–792, 1999.
21. Matveeva, O.V., Tsodikov, A.D., Giddings, M., Freier, S.M., Wyatt, J.R., Spiridonov, A.N., Shabalina, S.A., Gesteland, R.F., and Atkins, J.F., Identification of sequence motifs in oligonucleotides whose presence is correlated with antisense activity, *Nucl. Acids Res.*, 28, 2862–2865, 2000.
22. Wetmur, J.G., DNA probes: applications of the principles of nucleic acid hybridization, *Crit. Rev. Biochem. Molec. Biol.*, 26, 227–259, 1991.
23. Hughes, T.R., Mao, M. et al., Expression profiling using microarrays fabricated by an ink-jet oligonucleotide synthesizer, *Nat. Biotechnol.*, 19, 342–347, 2001.
24. Yu, H., Chao, J., Patek, D., Mujumdar, R., Mujumdar, S., and Waggoner, A.S., Cyanine dye dUTP analogs for enzymatic labeling of DNA probes, *Nucl. Acids Res.*, 22, 3226–3232, 1994.
25. Langer, P.R., Waldrop, A.A., and Ward, D.C., Enzymatic synthesis of biotin-labeled polynucleotides: novel nucleic acid affinity probes, *Proc. Natl. Acad. Sci. U.S.A.*, 78, 6633–6637, 1981.
26. Hastie, N.D. and Bishop, J.O., The expression of three abundance classes of messenger RNA in mouse tissues, *Cell*, 9, 761–774, 1976.
27. Lee, M.-L. T., Kuo, F.C., Whitmore, G.A., and Sklar, J., Importance of replication in microarray gene expression studies: statistical methods and evidence from repetitive cDNA hybridizations, *Proc. Natl. Acad. Sci. U.S.A.*, 97, 9834–9839, 2000.
28. Yue, H., Eastman, S.P. et al., An evaluation of the performance of cDNA microarrays for detecting changes in global mRNA expression, *Nucl. Acids Res.*, 29, e41, 2001.
29. Golub, T.R., Slonim, D.K. et al., Molecular classification of cancer: class discovery and class prediction by gene expression monitoring, *Science*, 286, 531–537, 1999.
30. Alizadeh, A.A., Eisen, M.B. et al., Distinct types of diffuse large B-cell lymphoma identified by gene expression profiling, *Nature*, 403, 503–511, 2000.
31. Welsh, J.B., Zarrinkar, P.P. et al., Analysis of gene expression profiles in normal and neoplastic ovarian tissue samples identifies candidate molecular markers of epithelial ovarian cancer, *Proc. Natl. Acad. Sci. U.S.A.*, 98, 1176–1181, 2001.
32. Clark, E.A., Golub, T.R., Lander, E.S., and Hynes, R.O., Genomic analysis of metastasis reveals an essential role for RhoC, *Nature*, 406, 532–535, 2000.
33. McCarthy, J.J. and Hilfiker, R., The use of single nucleotide polymorphism maps in pharmacogenomics, *Nat. Biotechnol.*, 18, 505–508, 2000.
34. Herdewijn, P., Targeting RNA with conformationally restricted oligonucleotides, *Liebigs Ann.*, 1337–1348, 1996.
35. Majlessi, M., Nelson, N.C., and Becker, M.M., Advantages of 2′-O-methyl oligoribonucleotide probes for detecting RNA targets, *Nucl. Acids Res.*, 26, 2224–2229, 1998.

4 Electrochemical Detection of Nucleic Acids

Allen Eckhardt, Eric Espenhahn, Mary Napier, Natasha Popovich, Holden Thorp, and Robert Witwer

CONTENTS

4.1 INTRODUCTION

XANTHON™, Inc., of Research Triangle Park, North Carolina, was founded in 1996 to commercialize a novel detection technology discovered at the University of North Carolina at Chapel Hill. Technology development efforts have culminated in a unique platform that unites microelectronics (semiconductor microcircuitry), molecular biology, and electrochemistry for the direct detection of nucleic acids. The method is based on the ability of a chemical mediator to exploit the naturally occurring guanine content of nucleic acids. In the XANTHON system, an electrical

0-8493-2285-5/02/$0.00+$1.50

potential is applied to an electrode in the presence of a transition-metal complex and a current is produced that is proportional to the amount of guanine present at the electrode. This reaction is quantitative and highly reproducible, offering direct detection of nucleic acids without the need for sample purification, sample amplification, or the introduction of chemical labels.

The initial platform for incorporation of the XANTHON technology is a 96-well microplate. Each well in the XANTHON plate contains nine electrodes: seven 200-μm "working" electrodes, a counter-electrode and a reference electrode. A different nucleic acid probe sequence is affixed to each "working" electrode within a well, enabling hybridization reactions to localize target nucleic acids at individual electrodes, thus providing simultaneous "multiplexed" detection of up to seven target nucleic acids per sample. To commercially implement this system, XANTHON has developed a benchtop instrument and customized data analysis software in addition to the disposable 96-well XANTHON plate.

The initial application of the XANTHON technology is gene expression analysis, that is, measurement of specific messenger RNA (mRNA) molecules. The unique features of the XANTHON electrochemical detection system make it an ideal tool for use by organizations involved in pharmaceutical drug discovery/development, specifically in secondary screening of compounds and lead compound optimization, by accurately and reliably assessing the impact of large numbers of compounds on the expression of a discrete bundle of genes in cell-based assays. The XANTHON system generates the same information as standard gene expression methods, but does it hundreds of times faster, and with less opportunity for error and at a lower cost.

This chapter introduces basic electrochemical detection technologies and focuses on the specific electrochemical techniques, immobilization methods, assay procedures, microfabrication processes, and instrumentation that, together, comprise the XANTHON electrochemical detection system.

4.2 BACKGROUND

The attachment of nucleic acids to spatially resolved microlocations in array formats has enabled new technologies for monitoring the presence of infectious organisms, quantitating mRNA expression levels, and sequencing genomic DNA.[1–3] Factors that have made these advancements possible include fluorescence microscopy, methods to label DNA and RNA with fluorophores, and the ability to direct nucleic acids to microlocations using either deposition[2] or photolithography.[3] Electrochemical techniques are ideally suited to miniaturization and have the potential to simplify nucleic acid analysis by circumventing the need for fluorescent labeling steps and fluorescent microscopy.[4–8]

A number of approaches have been used to realize direct electrochemistry of nucleic acids where electrons are exchanged directly between a solid electrode and the molecular components of DNA or RNA. Direct electrochemical reduction of nucleobases was first demonstrated by Palecek via adsorption of DNA or RNA onto mercury electrodes[9] and subsequent optimization has produced methods for detecting DNA with a sensitivity of roughly 100 pg in 5 μl.[10,11] Wang et al.[12–15] have applied a similar strategy to the detection of direct oxidation of the guanine and adenine bases adsorbed onto carbon electrodes. The major difficulty with this oxidative approach is that water is also oxidized at the potentials needed to oxidize guanine and adenine, although using background subtraction, Wang et al. have detected 15.4 fmol of tRNA at a 2.5-mm diameter carbon paste electrode.[15] Kuhr et al.[16–18] have also achieved a sensitive detection system based on catalytic oxidation of the deoxyribose (or ribose) sugar in nucleic acid at a copper microelectrode. In this system, sinusoidal voltammetry has been used to detect DNA concentrations of 3.2 p*M*.[17,18]

A difficulty with direct electrochemistry of nucleic acids is that the rates at which electrons are transferred from the nucleic acid to the electrode are often slow, decreasing the measured currents.[19] These poor electron-transfer kinetics have led to the development of systems in which detection of nucleic acids is coupled to the detection of some secondary reporter molecule. Mikkelsen and co-workers[20,21] have proposed a method where DNA is covalently coupled directly to carbon electrodes and enhanced faradaic current of $Co(bpy)_3^{2+/3+}$ is measured in the presence of the surface-attached DNA. A system recently reported by Heller et al.[22] involves modification of DNA with horseradish peroxidase to allow coupling to "wired" polymer electrode surfaces. The redox chemistry of methylene blue is also readily monitored on alkanethiol monolayers of DNA on gold electrodes and is sensitive to the DNA structure.[8,23]

XANTHON has created a sensitive nucleic acid detection system using $Ru(bpy)_3^{2+/3+}$ as a guanine oxidation catalyst (bpy = 2,2′-bipyridine).[6,24] The $Ru(bpy)_3^{3+}$ complex exhibits rapid electron-transfer kinetics with most electrode materials, including the tin-doped indium oxide (ITO) electrodes used in the XANTHON system.[25] The redox potential of $Ru(bpy)_3^{3+/2+}$ is 1.06 V vs. Ag/AgCl, which is close to that of the guanosine radical cation (E^0 = 1.07 V vs. Ag/AgCl).[26] Thus, $Ru(bpy)_3^{3+}$ undergoes a thermoneutral reaction with guanine to generate $Ru(bpy)_3^{2+}$ and an oxidized guanine. The second-order rate constants for electron transfer from guanine in DNA to $Ru(bpy)_3^{3+}$ are as high as 10^6 M^{-1} s^{-1}.[27–31] Therefore, when $Ru(bpy)_3^{3+}$ is generated at an electrode by oxidation of $Ru(bpy)_3^{2+}$, the $Ru(bpy)_3^{3+}$ is immediately consumed by guanine oxidation to regenerate $Ru(bpy)_3^{2+}$. When the oxidation of $Ru(bpy)_3^{2+}$ is observed by an electrochemical experiment such as cyclic voltammetry as shown in Figure 4.1, addition of DNA to the system causes an increase in the oxidation current because of the regeneration of $Ru(bpy)_3^{2+}$, which is reoxidized by the electrode:

$$Ru(bpy)_3^{2+} \rightarrow Ru(bpy)_3^{3+} + e^- \quad (4.1)$$

$$Ru(bpy)_3^{3+} + \text{Guanine} \rightarrow Ru(bpy)_3^{2+} + \text{Guanine}^+ \quad (4.2)$$

The system circumvents many of the difficulties described above because Equation (4.2) greatly speeds up the delivery of electrons from guanine to the electrode and because ITO electrodes exhibit little water oxidation current at the potentials needed to achieve electrocatalysis.[25]

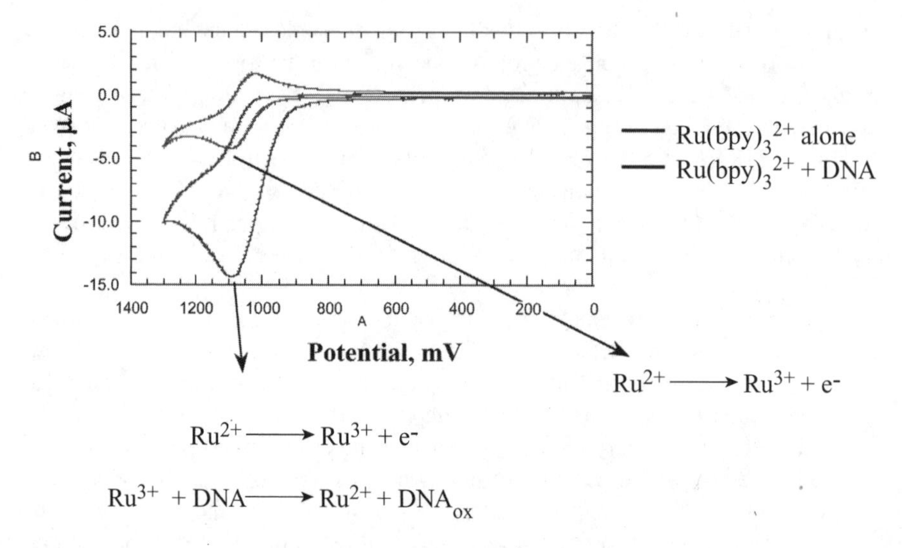

FIGURE 4.1 Cyclic voltammogram of $Ru(bpy)_3^{2+}$ with and without DNA.

To make this reaction scheme practical for use as a nucleic acid biosensor, surface immobilization of the nucleic acid probe on the electrode, as shown in Figure 4.2, must be achieved.[6] In the system, guanine-free probes can be prepared by substituting inosine for guanine.[32] If no target hybridizes to the probe, no catalytic current is observed upon interrogation of the electrode in the presence of $Ru(bpy)_3^{2+}$. However, when nucleic acid target hybridizes to complementary probe, then guanines present in the target are immobilized in close proximity to the electrode and catalytic current is observed upon interrogation of the electrode in the presence of $Ru(bpy)_3^{2+}$. When large targets (such as intact mRNA) are hybridized then very large current enhancements are observed upon oxidation of $Ru(bpy)_3^{2+}$ at the electrode surface.[33] A number of immobilization schemes have been utilized, including self-assembled monolayers,[33] electropolymerized films,[34] porous membranes attached to the ITO surface,[35] direct attachment to ITO electrodes,[19,36] and silane attachment.

4.3 ELECTROCHEMISTRY

Electrochemical methods are ideally suited for miniaturization. Microelectrodes offer higher sensitivity, smaller double-layer capacitance, and lower ohmic losses than macroelectrodes, resulting in a higher signal-to-noise (S/N) ratio. To implement the XANTHON technology in a 96-well microtiter plate format, it was necessary to reduce the size of the working electrode from the 6-mm diameter used in the experimental setting to generate preliminary data. A 200-µm electrode size was chosen because this feature size is easy to fabricate, and the currents obtained are in the nanoampere (nA) range, which can be readily measured using conventional electrochemical instrumentation. With the reduction in electrode size, sensitivity levels moved from µA/pmol guanine to nA/attomole guanine.

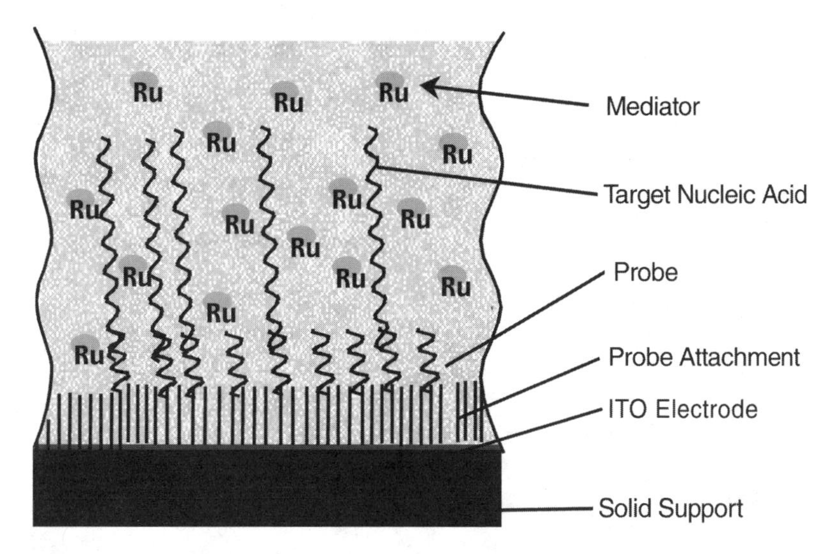

FIGURE 4.2 Diagram of electrode with attached probes and hybridized targets.

Cyclic voltammetry is one electrochemical method that can be employed for interrogation using the XANTHON technology. It is used for product development purposes because it provides information on the mechanism of the electrochemical reaction under study, as well as quantitative information. In this method, a potentiostat is used to apply a potential to the working electrode with respect to the reference electrode whose potential is well-defined and constant during the experiment. The potential of the working electrode is increased linearly with time to a specified value at a constant rate. When the specified value is reached, the potential is then reversed to the starting point at the same rate. This change in potential is the electrochemical driving force that causes oxidation or reduction of the analyte. The current resulting from these events is measured and recorded in a cyclic voltammogram, which is a plot of current as a function of the potential applied to the working electrode. Figure 4.3 shows representative cyclic voltammograms obtained at 200-μm electrodes. As the amount of guanine immobilized at the electrode surface increases, the peak current increases in a linear fashion.

When cyclic voltammetry is used, the amount of signal obtained is a function of scan rate. This limits the time spent at the peak potential and does not allow for complete oxidation of guanine residues present at the electrode surface. The XANTHON electrochemical detection system utilizes chronoamperometry, which permits collection of the maximum electrochemical signal from each guanine and is easier to implement as it involves only a single potential step. Allowing sufficient time for full oxidation to occur is essential because it increases the number of electrons collected per guanine and thus has a profound effect on the sensitivity of the method. The fact that there are multiple guanine residues in the target nucleic acid, and that each guanine provides multiple electrons to the electrode, results in amplification of the signal and accounts for the high sensitivity of this system. It has been reported

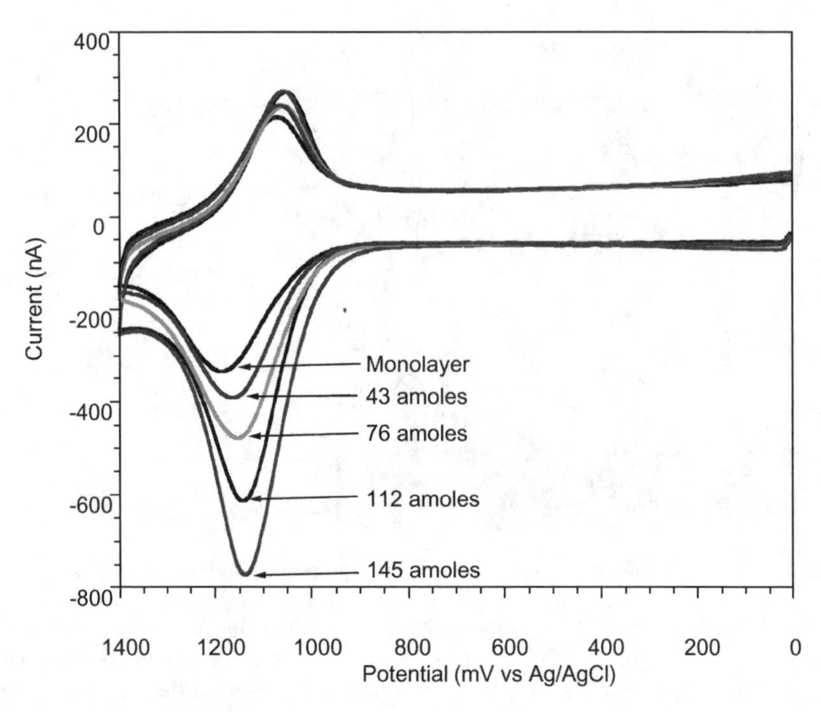

FIGURE 4.3 Dose response of a 21mer probe with five guanines immobilized on a 200-μm electrode.

that the number of electrons collected per guanine ranges from two to six, depending on the interrogation method employed and the solid phase used for immobilization.

In chronoamperometry, the potential of the working electrode is stepped from a value where no electron-transfer reaction occurs (e.g., 0 mV) to a potential where the mediator is oxidized (1100 mV) and able to extract electrons from guanine. The working electrode is held at the step potential for a specified period of time and current is measured as a function of time. The signal generated by the target nucleic acid is obtained by integrating the current measured during a specified time, resulting in charge passed as a result of guanine oxidation. This value is linearly proportional to the amount of guanine at the electrode surface. Current is integrated during the time window that maximizes the catalytic signal from guanine oxidation while minimizing the background signal. In the XANTHON system, each electrode is interrogated three times, permitting each electrode to serve as its own control. The first potential step results in complete oxidation of all guanine residues in the hybridized target strand, while the second and third potential steps allow for measurement of the pure diffusional current of $Ru(bpy)_3^{2+}$. Thus, the signal for the hybridized target is background-corrected for the diffusional signal on the same electrode so that minor differences in electrode area or solid-phase assembly are taken into account. This approach is superior to background correction using a separate control electrode and eases the demands on the electrode fabrication process.

The XANTHON system also utilizes a unique feature of electrochemical detection, namely the multiplexing capability that stems from the fact that the signal generated is localized to each electrode. When multiple electrodes are used, it is essential that there be no diffusional cross-talk between the electrodes. This issue can be avoided by providing adequate spacing between the electrodes. In the present layout, there is no cross-talk between the electrodes and the number of electrodes per well could be increased significantly. Theoretical calculations show that 30, 200-μm diameter electrodes can easily be placed at the bottom of a single well of a 96-well microtiter plate without diffusional cross-talk. In the first generation XANTHON plate, each well contains seven working electrodes plus the counter and reference electrodes necessary to conduct the electrochemical interrogation.

4.4 PROBE IMMOBILIZATION

The XANTHON technology relies on a sequence-specific capture probe for hybridization of the complementary target nucleic acid near the electrode surface. Hybridization of the target to the probe captures the target nucleic acid of interest from all other components in the sample mixture and brings the target in close proximity to the electrode surface for electrochemical analysis.

For probe immobilization, a biological layer is deposited onto the ITO working electrode surface. This biological layer consists of two components: the nucleic acid capture probe coupled to a molecule that anchors the capture probe to the electrode surface, and a molecule or combination of molecules that inhibits nonspecific binding of sample matrix components such as proteins, noncomplementary nucleic acids, carbohydrates, etc. to the electrode.

The biological layer has specific constraints for operation in the XANTHON system. Most importantly, the biological layer must be compatible with and stable under the conditions required for plate processing and target hybridization, as well as electrochemical interrogation. The biological layer must not contribute to background electrochemical signal or interfere with efficient electron transfer. Also, it must not nonspecifically attract sample matrix components to the electrode, which in turn could contribute background electrochemical signal or inhibit electron transfer. The biological layer must also make the immobilized capture probe available for efficient hybridization to the target nucleic acid and be compatible with the standard dispensing technologies used to deposit the layer onto the seven-electrode pattern in each well of the XANTHON plate.

Several strategies have been explored for immobilization of the biological layer at the working electrode. In one approach, self-assembled monolayers consisting of nucleic acid capture probes coupled to alkylphosphate or alkylphosphonate anchoring groups are utilized. This approach was chosen because phosphonate and phosphate anchoring groups both have a high affinity for metal oxide surfaces.[37] Carboxyl-terminated alkylphosphonate is combined with nucleic-acid-terminated alkylphosphonate or nucleic-acid-terminated alkylphosphate in a 10:1 ratio in DMSO/H_2O and applied to the ITO surface to form a mixed self-assembled monolayer on the electrode surface. Exposure of these monolayer systems to short synthetic nucleic acid targets results in hybridization of complementary target and reasonable discrimination against mismatch target.

In a second approach, silane anchoring molecules coupled to nucleic acid capture probes are utilized as the immobilization strategy. The silane anchor provides a robust attachment to the ITO surface; however, the reactivity of the silane moiety makes it difficult to devise a strategy for coupling of the nucleic acid without full hydrolysis and self-condensation of the silane prior to deposition on and attachment to the ITO electrode. One approach is to first derivatize the ITO electrode surface with a silane molecule having a highly reactive group capable of reacting with a modified nucleic acid. This approach is inefficient and yields a low density of probe on the electrode surface. The second approach is to couple reactive silanes to modified nucleic acid probes in solution, followed by immediate deposition onto the electrode surface. Because the reaction conditions in solution are much easier to control and monitor, this strategy allows better control of the biological layer and much higher capture probe density. A variety of silanes modified with N-hydroxysuccinimide (NHS) esters, epoxides, iodo, mercapto, succinic anhydride, and isocyanato chemistries were reacted with thiol and amine-terminated nucleic acid probes under various reaction conditions in solution that were tailored to the reactive groups on the silanes and on the nucleic acid probes. The reaction products were analyzed and the reaction mixtures deposited onto the electrode to determine which combination of reactive silane and modified nucleic acid would yield optimal results. Based on initial investigations, (isocyanatopropyl)triethoxysilane reacted with amine-terminated nucleic acid probes with high efficiency and resulted in an acceptable probe density on the ITO electrode surface.

In the current probe deposition protocol, (isocyanatopropyl)triethoxysilane is mixed in a 10:1 ratio with the amine-terminated probe in a basic pH buffer and allowed to react for 60 to 90 min. Using HPLC and mass spectrometry, this reaction was found to proceed to approximately 80% completion, with the major reaction product being the nucleic acid covalently coupled to the silane. The pH of the reaction mixture is then acidified to promote condensation of the silane on the ITO surface and deposited on the electrode using a commercially available small-volume deposition instrument. The deposited reaction mixture is placed in a humidified environment overnight, cured, and then washed with NaOH.

A schematic representation of the immobilized biological layer is shown in Figure 4.4. Coupling of a guanine-containing nucleic acid probe to the silane followed by deposition onto the ITO electrode results in an immobilized nucleic acid capture probe that can be followed electrochemically. Figure 4.5 illustrates the increase in signal due to the metal-mediated electron transfer from the guanine base of the nucleic acid probe to 200-μm ITO electrodes observed following deposition of a 13mer probe containing four guanines. The hybridization capability of the immobilized capture probe was analyzed using short synthetic nucleic acid match and mismatch targets. A 13mer capture probe containing no guanines was immobilized on 6-mm ITO electrodes and then exposed to either radiolabled 49mer match target with 17 guanines or 45mer mismatch target with 14 guanines. As shown in Figure 4.6, discrimination by both radiolabel and electrochemical methods was observed in the system.

$(Et_2O)_3Si$ ⁄\⁄\ $N{=}C{=}O$ + NA —$NHCO(linker)NH_2$

↓ Allow to react for 90 minutes at basic pH

$(Et_2O)_3Si$ ⁄\⁄\ NHCONH(linker)CONH-NA

↓ Spot onto electrode in acid buffer
Cure at elevated temperature

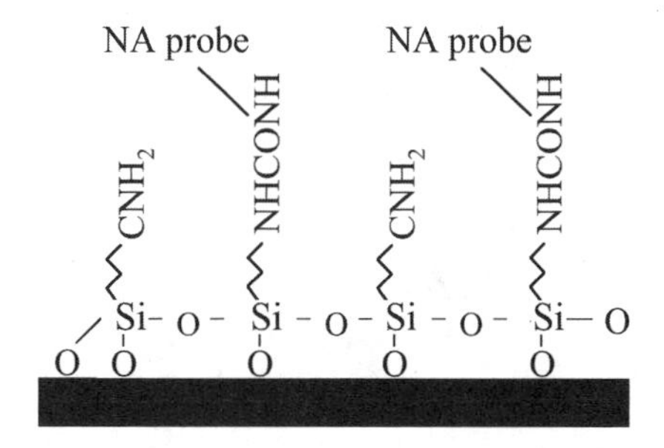

FIGURE 4.4 Schematic representation of a silane-anchored nucleic acid capture probe.

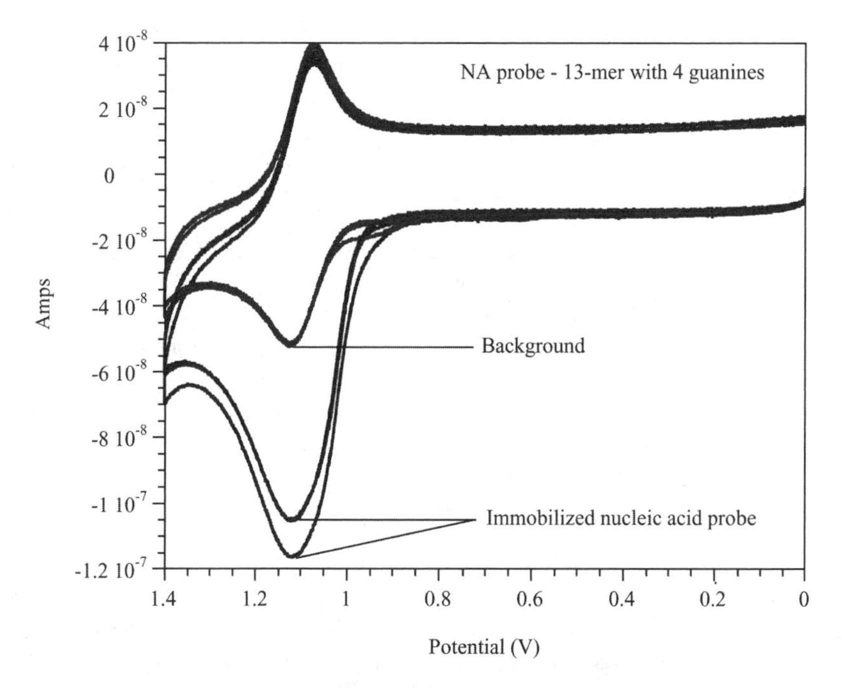

FIGURE 4.5 Electrochemical signal generated from immobilized capture probe containing four guanines. The increase in signal over background is due to the metal-mediated electron transfer from the guanine base of the probe to the 200-μm ITO electrodes.

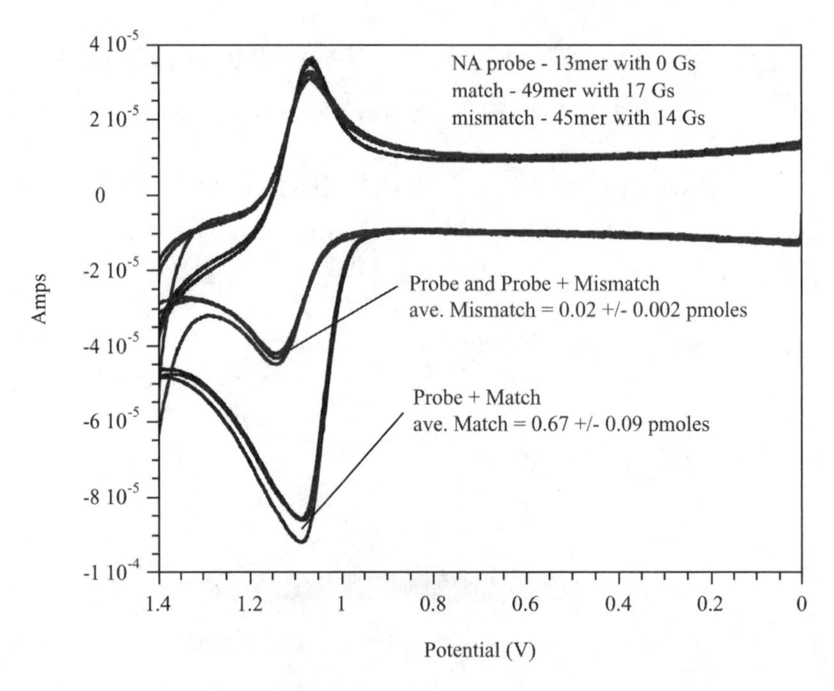

FIGURE 4.6 Electrochemical signal generated from synthetic nucleic acid target hybridized to immobilized capture probe on 6-mm ITO electrodes. The capture probe is a 13mer with no guanines. The match target is a 49mer containing 17 guanines and the mismatch target is a 45mer containing 14 guanines. No increase in signal was observed for the electrodes exposed to the mismatch target.

4.5 MOLECULAR BIOLOGY

The XANTHON system is designed to measure expression levels of messenger RNA (mRNA) in a quantitative manner using cell-based assays in a 96-well microtiter plate format. This system allows for the accurate measurement of the induction or repression of selected genes in response to chemical or physiological treatment of cultured cells in a high-throughput format. The assay does not require purification, reverse transcription, amplification, or external labeling of the mRNA. Messenger RNA detection is direct and is solely dependent on the number of guanines at the electrode surface. Factors that determine the number of guanines at the electrode surface include the length of the mRNA, its guanine content, and the number of mRNA molecules hybridized at the electrode.

The XANTHON system is a hybridization-based detection system. This means nucleic acid hybridization is used to increase both the specificity and sensitivity of the assay. Hybridization is used to select specific mRNA molecules from complex mixtures of total RNA by hybridization of the desired mRNA target to a complementary nucleic acid probe immobilized on an electrode. Hybridization also increases sensitivity by effectively increasing the concentration of the target mRNA at the electrode surface, which leads to an enhancement in the chronoamperometric signal.

As mentioned, each well of the microtiter plate contains seven working electrodes and each electrode has a different immobilized capture probe. This layout allows for simultaneous multiplexing or the determination of the expression level of seven different mRNA targets in the same sample. A possible assay format that includes appropriate controls would use five electrodes for measuring the expression of five different mRNA targets while one electrode would serve as a positive control and the remaining electrode would serve as a negative control. The positive control would be a capture probe complementary to a housekeeping gene, and the signal from this electrode would be used to perform sample-to-sample normalization. The negative control would be a capture probe that is noncomplementary to any possible target in the cell line being tested. The signal from this electrode would be used to monitor for any adverse event, such as cellular components or debris reacting with or precipitating onto the electrodes, incomplete washing of the well, or the presence of redox-active test chemicals that interact with the electrodes and produce an electrochemical signal upon interrogation.

Selection of capture probe sequences is of utmost importance in achieving assay specificity and in maximizing assay performance. The probe must be of sufficient length and contain a complementary sequence with unique specificity to ensure hybridization to only one mRNA target. We have successfully used probes with lengths ranging from 13 to 25 nucleotides. Capture probe sequences should also be selected to complementary regions in the target molecule that have zero or very few cytidines. This minimizes the number of guanines in the capture probe, thereby keeping the background signal to an absolute minimum. As needed, electrochemically silent guanine analogs such as inosine can be substituted in the capture probe, which will reduce the background signal during interrogation.

The capture probes can be composed of DNA, RNA, or nucleic acid analogs such as PNA, that perform satisfactorily in the system. Another important parameter is the density of the capture probes attached to the electrodes. To achieve maximum hybridization of target mRNA and optimal electrochemical signal, the amount of probe on a 200-μm electrode should be between 3 and 10 fmol. Also, spacer molecules are inserted between the capture probe and the attachment chemistry used to immobilize the probe on the electrode. The composition and length of these spacers are critical to maximize target hybridization and electrochemical signal. The optimum spacer to date is composed of polyethylene and glutamic acid groups that extend the capture probe approximately 45 Å above the ITO surface.

One of the features of the XANTHON system is that it will detect mRNA targets in crude cell lysates with minimal sample preparation. To achieve this goal, one of the critical components of the system is the cell lysis solution. This solution must accomplish several tasks. There must be complete cell lysis and solubilization of nucleic acids; but given that target integrity is so important, there can be no enzymatic degradation or mechanical shearing of the mRNA. The cell lysis solution must completely inactivate all ribonucleases while simultaneously disrupting the mRNA secondary structure to expose complementary hybridization sites to allow efficient hybridization of the target mRNA to the capture probe. Finally, components of the cell lysis solution must be electrochemically inactive and must not bind to or interfere with the working, reference, or counter electrodes within the microtiter well.

A related issue of particular concern is the nonspecific binding of cellular components such as nucleic acids and proteins to the electrodes, the passivation layer, or the polystyrene plate upper. Nonspecific binding can potentially foul electrodes, and interfere with the electrochemistry by blocking the electrodes or contributing additional unwanted signal. The binding of nucleic acids to the passivation layer or the plate upper will divert nucleic acid away from the electrodes, thereby reducing the signal for specific targets. To minimize nonspecific binding, the passivation layer is composed of a material that is extremely inert toward the adsorption of protein and nucleic acid. In addition, backcoating with nonproteinaceous materials has been used to decrease nonspecific binding to the absolute minimum. Typical backcoating materials that have proven effective in reducing nonspecific adsorption in this system include alkyl phosphonates, polyvinyl alcohol, polyinosine, and polyglutamic acid.

The assay for measuring mRNA expression levels in adherent cultured cells with the XANTHON system is very simple. In general, the cell culture medium is removed and the cells are washed with an isotonic phosphate buffered saline. The cells are lysed by adding 50 μl of cell lysis buffer and incubating for 15 min with mild agitation. This volume of cell lysis buffer is sufficient to solubilize 50,000 HepG2 cells in one well of a microtiter plate. The entire cell lysate is then transferred to one well of a XANTHON plate and allowed to hybridize overnight. After hybridization, the cell lysates are removed, the plates are washed, $Ru(bpy)_3^{2+}$ mediator is added, and the plates are electrochemically interrogated.

To optimize the system, we have used both synthetic oligonucleotides and *in vitro* transcribed mRNA targets diluted into either aqueous buffer or crude HepG2 cell lysate. The matching RNA target in the model system is apolipoprotein A1 (Apo A1, 973 nucleotides, 305 guanines) and the mismatch RNA target is alpha-1-microglobulin/bikunin precursor (AMBP, 1034 nucleotides, 275 guanines). Figure 4.7 shows cyclic voltammograms from 200-μm electrodes after hybridization of apolipoprotein A1 or AMBP mRNA in HepG2 cell lysates. The capture probe is a 13mer guanine-free sequence that is complementary to apolipoprotein A1. As seen in this figure, hybridization of complementary target in the crude cell lysate led to a large increase in current when compared to the current generated from noncomplementary target. Extrapolating the data from the model system, the sensitivity of the assay will be approximately two to three copies per cell for a specific mRNA assuming a length of 1500 nucleotides and 50,000 cells per well. The assay will be able to detect relative changes in expression level for low, medium, and high copy number mRNA species over a range of four logs concentration with a minimum twofold detection limit difference. Ultra-high copy number species may require sample dilution so as not to saturate the system and exceed the maximum range of the instrument.

4.6 MICROFABRICATION

XANTHON has selected the well-known, laboratory standard 96-well footprint as the presentation platform for the multiplexed electrode configuration that is the heart of the XANTHON system. Although the plate footprint lends itself to easy integration into current laboratory protocols and use with robotic systems, it is quite unique in its architecture.

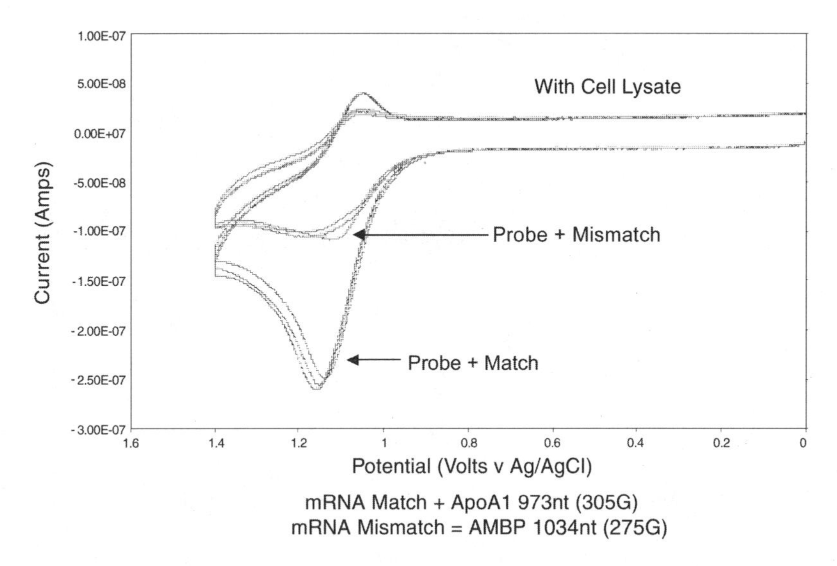

FIGURE 4.7 Electrochemical signal generated from Apo A1 mRNA target hybridized to an immobilized complementary 13mer capture probe with no guanines. The mismatch target is AMBP mRNA.

Each of the 96 wells contains a pattern of nine electrodes, including seven working electrodes, one reference electrode, and one counter electrode. As shown in Figure 4.8, the working and counter electrodes are individually connected to contact pads patterned on the backside of the plate through precision metal-filled pores or "vias" that have been drilled through the base material. This plate architecture mates with the conductor bed in the XANTHON instrument to complete the electrical circuit. Fabrication of the XANTHON plate is a convergence of highly technical disciplines, including semiconductor microcircuitry, molecular biology, and electrochemistry. This section provides an overview of the materials requirements, manufacturing processes, controls, and facility requirements for production of the XANTHON plates.

4.6.1 Material Requirements

The base of the plate is fabricated from a strong, rigid material, such as quartz or alumina, into which pores or vias can be formed and filled with a conductive material. The base also supports formation of circuit traces on both the top and bottom surfaces by permitting deposition of metals onto the surfaces and subsequent etching away of extraneous metal to form specific trace patterns. The base is flat to allow leak-free bonding to the plastic welled upper, and the uninterrupted mating of the bottom surface to the conductor bed in the instrument. The top surface finish is glass-like to facilitate fabrication of the electrodes. Finally, the base material is able to withstand process temperatures of up to 1400°C without losing form or shape.

The welled upper is a lightweight plastic that is injection molded to form the walls of the 96-well plate. It is bonded to the base with an electrochemically inert

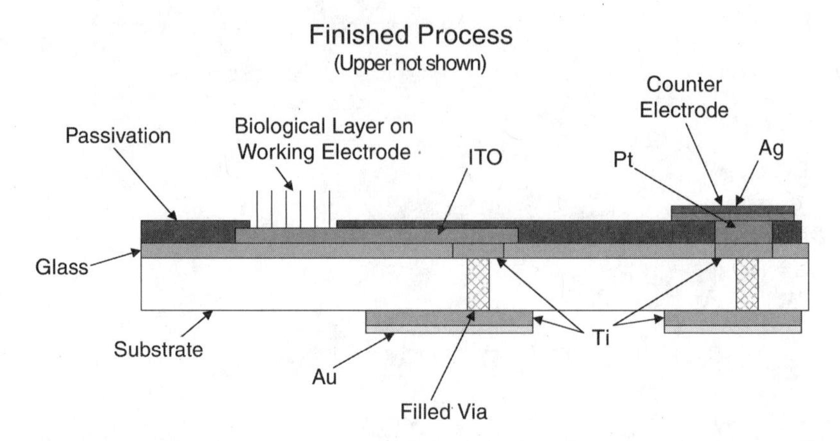

FIGURE 4.8 Cross-sectional view of the XANTHON plate base through one working electrode and the counter electrode showing various metalization and passivation layers.

adhesive and withstands temperatures up to 90°C. The upper must not attract biological materials or must be treated so as to prevent nonspecific binding and not interfere with the assay, the electrochemistry, or the bonding process.

The material of choice for the working electrodes, as determined by electrochemical reaction efficiency, is tin-doped indium oxide (ITO), which is deposited onto the base material using vacuum sputtering techniques. Other precious metals such as platinum, titanium, silver, and gold are applied to the base by vacuum sputtering or evaporation techniques to complete fabrication of the plate's conductive elements.

4.6.2 Processing

To maximize the efficiency of plate manufacturing and in-process control, the starting material is a substrate that will yield two base plates and multiple test "coupons." The test coupons provide an effective means for conducting destructive in-process testing and provide a history of physical processing without reducing the yield of usable bases. The virgin substrate is formed to size and vias are created. Each via is then filled with a material that forms a solid conductive path through the substrate from top to bottom. The substrate surfaces are then lapped and polished, as needed, to produce the glass-like surface required for electrode fabrication.

The process of fabricating the conductive elements begins by depositing multiple layers of metals onto each side of the substrate using vacuum processing. Electrodes and contacts are formed using standard semiconductor/thin-film type deposition and photolithography processes.

The normal sequence for photolithography is to spin-coat the surface with photoresist and expose it to ultraviolet light through a photomask that defines the final pattern of the electrical element. After the photoresist is exposed, developed, and microscopically inspected, the remaining photoresist is developed followed by etching of the underlying metal layer(s). The patterned photoresist is finally removed or stripped off, leaving the desired metal pattern. This process is repeated for each photo layer until the desired contact pad and trace patterns remain.

After the contact and electrode traces are completed, a passivation layer is applied over the entire surface and the seven working electrodes are formed by removal of the passivation material from the surface of the electrodes. As shown in Figure 4.8, the base subassembly is completed by depositing the counter and reference electrodes onto the passivation layer where they contact the underlying traces through openings in the layer. The substrate is then cut into its final dimensions using a diamond saw to yield two plate bases. Figure 4.9 shows the circuit pattern of the nine electrodes in one well.

Each base is bonded to the plastic welled upper using a screen-printing process in which adhesive is screened onto the base, the upper positioned, and the adhesive cured. Care must be taken to not allow adhesive into the well area and to ensure that the adhesive forms an integral seal to prevent leaks between wells.

Customer-specified probes are coupled to the attachment chemistry, suspended in a compatible solvent system, and deposited directly onto each working electrode using automated dispensing equipment. This equipment reproducibly locates and dispenses less than 10 nl of probe solution onto each working electrode.

4.6.3 Facilities

Most of the substrate processing is accomplished in a controlled environment or cleanroom. The photolithography process is accomplished under the most stringently controlled conditions, as it is extremely sensitive to particulate contamination. Dust and other particles can cause defects in the developed patterns that could result in

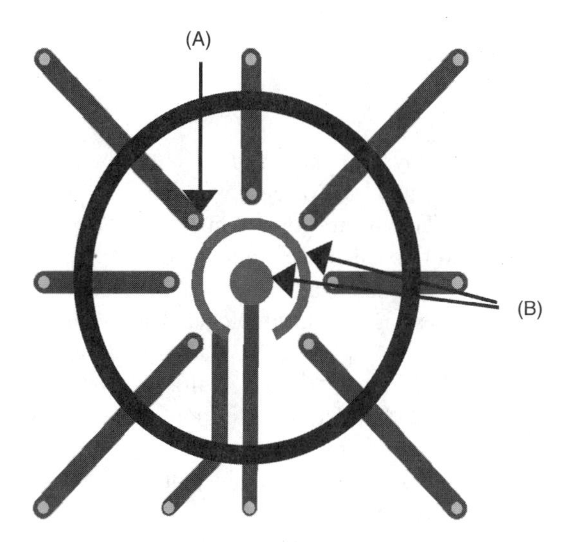

FIGURE 4.9 Top view of the integrated, multiplexed electrodes within one well of the XANTHON 96-well plate.

the formation of pinholes in the traces or passivation layers. The electrode deposition process is slightly more forgiving, but is still completed in a cleanroom environment.

4.6.4 Controls

Although the initial XANTHON product does not require FDA or ISO9000 certifications, all processes are being documented and validated to conform to these requirements. Design Control procedures and documentation have been established from the beginning of the project. As the product transitions from design through development to production, more extensive documentation is generated and incorporated into the manufacturing processes. Standard Operating Procedures guide all operations performed in the manufacturing processes.

As mentioned previously, each substrate has two integrated test coupons. One is an area that allows space for deposition of the metals so that thickness and adhesion can be monitored. The other is a series of eight sets of nine electrodes that can be used after the base has been fabricated to test and verify probe deposition and function. The base and each of the associated test coupons are marked with the same serial number so that they can be tracked throughout the manufacturing process, even after they have been separated.

Each process in the manufacturing sequence is monitored using statistical process control methods where appropriate. As with any process, monitoring trends and applying appropriate corrective actions before a problem occurs is foremost in controlling the fabrication process.

4.7 THE INSTRUMENT

4.7.1 Instrument Architecture

The XANTHON system quickly, accurately, and reproducibly measures target nucleic acids hybridized to probes immobilized on each working electrode in the multiplexed, 96-well XANTHON plate. The integrated system fits on a laboratory benchtop and requires only standard power connections. Figure 4.10 is a photograph of the XANTHON instrument showing the plate reader, plate handler, and computer modules.

The interrogation process is rapid, simple, and automated to accommodate high-throughput applications. Samples are prepared as described previously and transferred to the XANTHON plate for interrogation by the instrument. Plates are stacked in the input tower of the plate handler, an integrated Zymark Twister™. Each plate is moved, in turn, to the plate reader where each of the 96 wells is interrogated. After all the wells are processed, the plate is moved to the output tower of the plate handler. This cycle is repeated until the plate handler is empty, or the operator presses the *Stop* button on the instrument control computer (ICC), an IBM NetVista™. The operator can then review results in any of several formats on the instrument's computer or upload the data to a network server.

The XANTHON instrument interrogates each well in a plate using chronoamperometry. The electrical currents from each of the seven working electrodes are measured in parallel. Of these seven working electrodes, five contain target-specific

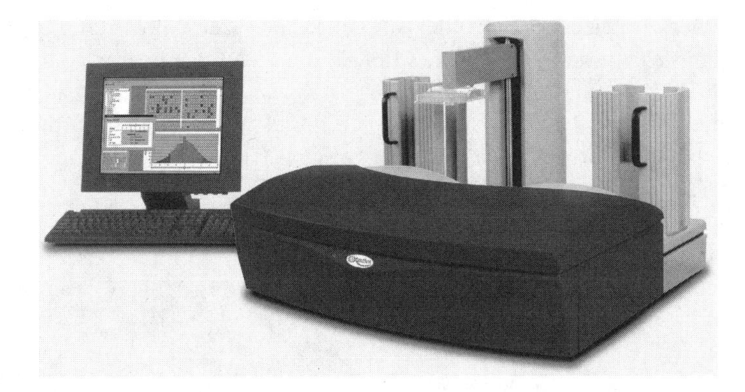

FIGURE 4.10 A photograph of the XANTHON instrument, with plate reader, plate handler, and instrument control computer (ICC).

probes, one is a negative control, and one is a positive control; therefore, 672 data points (7×96) are collected from each plate in less than 5 min.

A block diagram of the instrument is shown in Figure 4.11. The ICC runs the various subsystems of the instrument. It collects the data from the plate reader, performs signal processing, and stores the results so they can be analyzed later.

The plate reader controller is housed in the plate reader, where it coordinates the activities of this subsystem and provides communication to the ICC. It is a single-board computer that directly controls the motors, sensors, and other electronics in the plate reader.

The potentiostat processes one well at a time. It drives the counter electrode in such a way as to keep the reference electrode at a predefined potential relative to

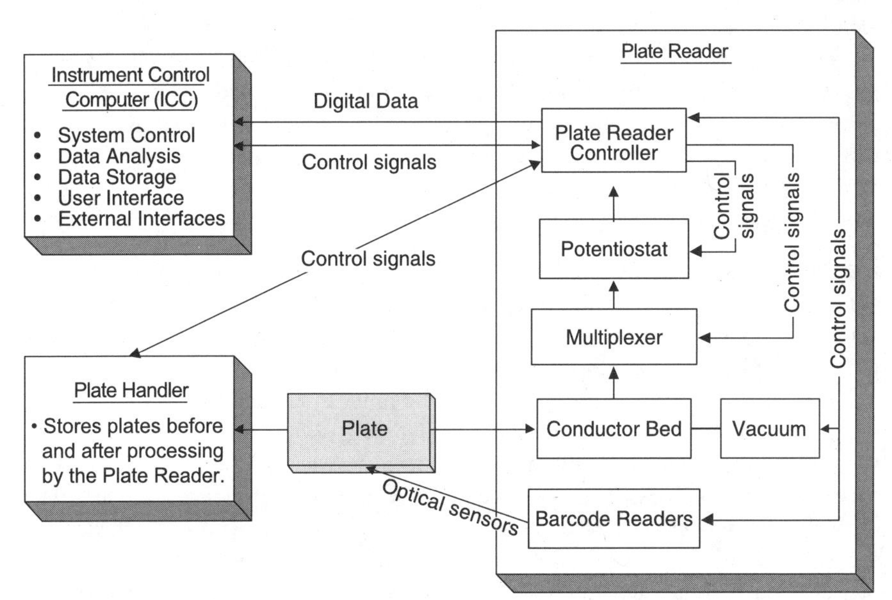

FIGURE 4.11 Block diagram of the instrument.

the seven working electrodes. At the same time, the electrical current generated at each working electrode is measured. After all the data for a well is collected, it is sent to the ICC for analysis and storage.

The multiplexer is used to switch the electrical connections from the selected well to the potentiostat. Using an electrical switch is more reliable than mechanically switching from one well to another.

The conductor bed shown in Figure 4.12 makes electrical contact between the plate and the multiplexer. Alignment pins in the conductor bed ensure that the plate is properly oriented. Once the plate is seated on the conductor bed, all the electrodes in the wells are mechanically connected, although not all electrodes are electrically active at the same time. The plate is held onto the conductors by a vacuum system that applies uniform pressure across the plate to ensure firm electrical connections.

FIGURE 4.12 Drawing of the conductor bed.

Bar code readers read the two bar code labels on the plate. One label, applied by XANTHON at the time of manufacture, contains a number that identifies the probes on the plate. The other label, applied by the customer, identifies the samples in the plate.

The plate handler has input and output towers and a robotic arm that moves the plates between the towers and the plate reader. Plates are queued in the plate handler before and after processing by the plate reader, allowing the instrument to run unattended. For low-volume applications, plates can be loaded manually.

4.7.2 Instrument Operation

After the prepared XANTHON plates are loaded into the input tower, the *Start* button on the ICC is selected. To account for environmental changes, the instrument automatically calibrates the electronics using extremely stable internal standards before any plates are processed.

The robotic arm removes a plate from the input tower and moves it to the plate reader. The plate is then set into the conductor bed of the plate reader where the vacuum system is engaged and the plate is firmly seated on the conductor bed.

Each well is processed in turn. The multiplexer circuit in the plate reader selects the well to be processed. The potentiostat then introduces an electrical potential into the well that initiates the electrochemical reaction. The potentiostat simultaneously converts the currents generated at the seven working electrodes to digital values. This will continue for a predefined amount of time, usually less than 100 ms.

After all the wells have been processed once (the first pass), the instrument repeats this process two more times. Because all of the guanines have been oxidized in the first pass, the second pass is used to calculate the "background" charge generated by the soluble mediator. After subtracting the background determined by the second pass from the first pass, the resulting charge is directly proportional to the number of guanines present in the target nucleic acid hybridized on the electrode. A third pass is performed to ensure that all guanines were oxidized during the first pass; therefore, results from the third pass should be similar to those of the second pass because they are both measuring background charge. Once all the electrical current measurements are made, the electrical charge is calculated by integrating the electrical current over time. A typical result for a single electrode showing the area integrated is presented in Figure 4.13.

4.7.3 Data Analysis

Results can be normalized once all the data is collected. The first step is to divide the electrical charge from a working electrode by the charge from the positive control electrode in the same well in order to adjust for variations in the number of cells in each well. The second step is to divide the signal from an electrode in a well that contained challenged cells by the signal from the corresponding electrode in a well with unchallenged cells. This number is biologically relevant because it represents the increase or decrease in expression of challenged cells compared to unchallenged cells.

The XANTHON software allows the user to analyze data generated by several different means. The software can be used to answer questions, find patterns, and detect trends. As shown in Figure 4.14, data can be viewed as a 96-well plate, as a histogram, as a barchart ordered by response, or as a table. Trends can be viewed across wells, across changes in compound properties, or across experiments. If a particular response pattern in mRNA expression is expected, the data can be searched and filtered according to user-defined criteria. Replicates (within and between plates) and control wells are also supported. All results generated by the system can be easily transferred to other data analysis tools.

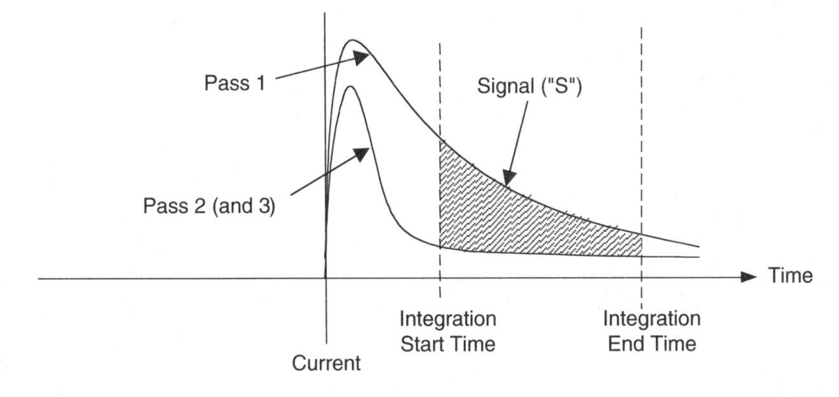

FIGURE 4.13 Graph showing calculation of the electrical charge.

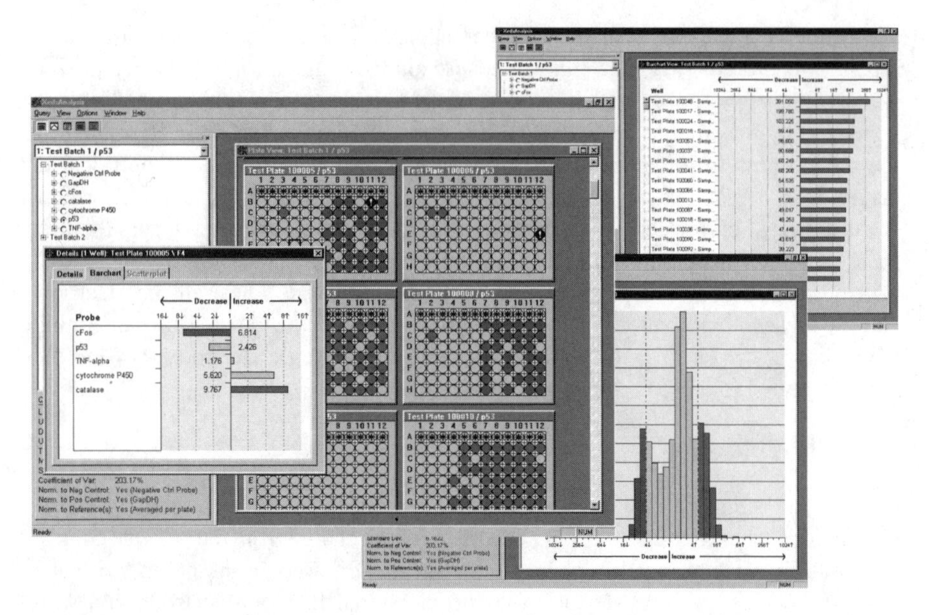

FIGURE 4.14 Instrument display screen showing results from one well. (See Color Figure 4.14 following page 82.)

REFERENCES

1. Spargo, C. A. et al., Chemiluminescent detection of strand displacement amplified DNA from species comprising the *Mycobacterium tuberculosis* complex, *Mol. Cell Probes*, 7, 395, 1993.
2. Schena, M. et al., Quantitative monitoring of gene expression patterns with a complementary DNA microarray, *Science*, 270, 467, 1995.
3. Chee, M. et al., Accessing genetic information with high-density DNA arrays, *Science*, 274, 610, 1996.
4. de Lumley-Woodyear, T., Campbell, C. N., and Heller, A., Direct enzyme-amplified electrical recognition of a 30-base model oligonucleotide, *J. Am. Chem. Soc.*, 118, 5504, 1996.
5. Mikkelsen, S. R., Electrochemical biosensors for DNA sequence detection, *Electroanalysis*, 8, 15, 1996.
6. Thorp, H. H., Cutting out the middleman: DNA biosensors based on electrochemical oxidation, *Trends Biotechnol.*, 16, 117, 1998.
7. Creager, S. et al., Electron transfer at electrodes through conjugated "molecular wire" bridges, *J. Am. Chem. Soc.*, 121, 1059, 1999.
8. Kelley, S. O. et al., Single-base mismatch detection based on charge transduction through DNA, *Nucleic Acids Res.*, 27, 4830, 1999.
9. Palecek, E., Oscillographic polarography of highly polymerized deoxyribonucleic acid, *Nature*, 188, 656, 1960.
10. Palecek, E. and Fojta, M., Differential pulse voltammetric determination of RNA at the picomole level in the presence of DNA and nucleic acid components, *Anal. Chem.*, 66, 1566, 1994.

11. Palecek, E. et al., Chronopotentiometric stripping of DNA at mercury electrodes, *Electroanalysis*, 9, 990, 1997.
12. Wang, J. et al., Adsorptive stripping potentiometry of DNA at electrochemically pretreated carbon paste electrodes, *Electroanalysis*, 8, 20, 1996.
13. Wang, J. et al., DNA electrochemical biosensor for the detection of short DNA sequences related to the human immunodeficiency virus, *Anal. Chem.*, 68, 2629, 1996.
14. Wang, J., Fernandes, J. R., and Kubota, L. T., Polishable and renewable DNA hybridization biosensors, *Anal. Chem.*, 70, 3699, 1998.
15. Wang, J. et al., Ultratrace measurements of nucleic acids by baseline-corrected adsorptive stripping square wave voltammetry, *Anal. Chem.*, 71, 1910, 1999.
16. Singhal, P. et al., Sinusoidal voltammetry for the analysis of carbohydrates at copper electrodes, *Anal. Chem.*, 69, 1662, 1997.
17. Singhal, P. and Kuhr, W. G., Ultrasensitive voltammetric detection of underivatized oligonucleotides and DNA, *Anal. Chem.*, 69, 4828, 1997.
18. Singhal, P. and Kuhr, W. G., Direct electrochemical detection of purine- and pyrimidine-based nucleotides with sinusoidal voltammetry, *Anal. Chem.*, 69, 3552, 1997.
19. Armistead, P. M. and Thorp, H. H., Oxidation kinetics of guanine in DNA molecules adsorbed onto indium tin oxide electrodes, *Anal. Chem.*, 73, 558, 2001.
20. Millan, K. M. and Mikkelsen, S. R., Sequence-selective biosensor for DNA based on electroactive hybridization indicators, *Anal. Chem.*, 65, 2317, 1993.
21. Millan, K. M., Saraullo, A., and Mikkelsen, S. R., Voltammetric DNA biosensor for cystic fibrosis based on a modified carbon paste electrode, *Anal. Chem.*, 66, 2943, 1994.
22. de Lumley-Woodyear, T. et al., Rapid amperometric verification of PCR amplification of DNA, *Anal. Chem.*, 71, 535, 1999.
23. Kelley, S. O. et al., Electrochemistry of methylene blue bound to a DNA-modified electrode, *Bioconjug. Chem.*, 8, 31, 1997.
24. Johnston, D. H., Welch, T. W., and Thorp, H. H., Electrochemically activated nucleic acid oxidation, *Met. Ions Biol. Syst.*, 33, 297, 1996.
25. Johnston, D., Glasgow, K., and Thorp, H., Electrochemical measurement of the solvent accessibility of nucleobases using electron transfer between DNA and metal complex, *J. Am. Chem. Soc.*, 117, 8933, 1995.
26. Steenken, S. and Jovanovic, S. V., How easily oxidizable is DNA? One-electron reduction potentials of adenosine and guanosine radicals in aqueous solution, *J. Am. Chem. Soc.*, 119, 617, 1997.
27. Johnston, D. H. and Thorp, H. H., Cyclic voltammetry studies of polynucleotide binding and oxidation by metal complexes: homogeneous electron-transfer kinetics, *J. Phys. Chem.*, 100, 13837, 1996.
28. Ropp, P. A. and Thorp, H. H., Site-selective electron transfer from purines to electrocatalysts: voltammetric detection of a biologically relevant deletion in hybridized DNA duplexes, *Chem. Biol.*, 6, 599, 1999.
29. Sistare, M. F., Holmberg, R. C., and Thorp, H. H., Electrochemical studies of polynucleotide binding and oxidation by metal complexes: effects of scan rate, concentration, and sequence, *J. Phys. Chem. B*, 103, 10718, 1999.
30. Szalai, V. A. and Thorp, H. H., Electron transfer in tetrads: adjacent guanines are not hole traps in G quartets, *J. Am. Chem. Soc.*, 122, 4524, 2000.
31. Weatherly, S. C., Yang, I. V., and Thorp, H. H., Proton-coupled electron transfer in duplex DNA: driving force dependence and isotope effects on electrocatalytic oxidation of guanine, *J. Am. Chem. Soc.*, 123, 1236, 2001.
32. Napier, M. E. et al., Probing biomolecule recognition with electron transfer: electrochemical sensors for DNA hybridization, *Bioconjug. Chem.*, 8, 906, 1997.

33. Napier, M. E. and Thorp, H. H., Modification of electrodes with dicarboxylate self-assembled monolayers for attachment and detection of nucleic acids, *Langmuir*, 13, 6342, 1997.
34. Ontko, A. C. et al., Electrochemical detection of single-stranded DNA using polymer-modified electrodes, *Inorg. Chem.*, 38, 1842, 1999.
35. Napier, M. E. and Thorp, H. H., Electrocatalytic oxidation of nucleic acids at electrodes modified with nylon and nitrocelluose membranes, *J. Fluorescence*, 9, 181, 1999.
36. Armistead, P. M. and Thorp, H. H., Modification of indium tin oxide electrodes with nucleic acids: detection of attomole quantities of immobilized DNA by electrocatalysis, *Anal. Chem.*, 72, 3764, 2000.
37. Gao, W. et al., Self-assembled monolayers of alkylphosphonic acids on metal oxides, *Langmuir*, 12, 6429, 1996.

5 DNA Microarrays in Neurobiology

Marie-Claude Potier, Geoffroy Golfier, Bruno Cauli, Natalie Gibelin, Beatrice Le Bourdelles, Bertrand Lambolez, Sonia Kuhlmann, Philippe Marc, Frédéric Devaux, and Jean Rossier

CONTENTS

5.1 INTRODUCTION

DNA microarrays are used to measure the expression patterns of thousands of genes in parallel, providing a sensitive, global readout of the physiological state of a cell or tissue samples and generating clues to gene functions. They are also used to monitor changes in gene expression in response to any physiological, pathological, or pharmacological variations.

We are using the DNA microarray technology on microscope slides developed by Patrick Brown at Stanford University.[1,2] DNA molecules are spotted on polylysine-coated slides. Two spotting machines are routinely used: one, making

0-8493-2285-5/02/$0.00+$1.50

spots by contact with glass (Genemachine, CA) with MicroQuill 1000 (Majer Precision, Inc.); and another from Gesim (Germany) with a piezoelectric device included inside a picoliter pipette. After hybridization of the microarrays with fluorescent probes (PCR products, cDNAs), the analysis of the slides is performed on a scanner from General Scanning capable of analyzing Cy3 and Cy5 fluorescence on spots at a resolution of 10-μm per pixel. The scanner is connected to a computer with image software analysis: Imagene 2 and 3 (Biodiscovery, Inc., Los Angeles, California) and Scanalyze 2.32.[3] Steady-state mRNA levels are deduced from the fluorescence intensity at each position on the microarray. With this technology and this equipment, it is possible to produce DNA microarrays of low and high density (hundreds to thousands and tens of thousands spots) on up to 100 slides in the same spotting experiment.

The mammalian nervous system is composed of a complex mosaic of billions of extremely diverse neurons (approximately 10^{11} neurons in the human brain). This large neuronal diversity is mainly generated by the differential expression of numerous genes. With the very high degree of complexity of the brain, the DNA microarray technology applied at the single-cell level should help in classifying cells according to their gene contents and define new targets for therapeutic intervention. By correlating gene expression patterns with specific physiological properties, it is possible to gain mechanistic insight into a broad range of neurobiological processes. The single-cell Reverse Transcription (RT)-PCR was developed 10 years ago[4] and it is based on the combination of the patch-clamp technique with RT-PCR. The electrical properties and the expression of different homologous genes are analyzed from a visually identified neuron. At the end of the patch-clamp recording, the cytoplasm of the neuron is harvested under visual and electrophysiological controls into the recording pipette by applying a gentle negative pressure. The content of the pipette is then expelled into a test tube and subsequently submitted to RT-PCR.

Recently, the amplification procedure of the single-cell RT-PCR was slightly modified in order to detect the expression of multiple nonhomologous genes.[5–10] This technique, called a multiplex RT-PCR, uses a specific set of PCR primers for each gene to be amplified. Using this approach, it is now possible to detect simultaneously the expression of 100 genes at the single-cell level. Here, we report the application of DNA microarrays for analysis of the expression of 100 genes in a single cell and in a whole tissue (brain extracts).

5.2 MATERIALS AND METHODS

5.2.1 Solutions

Diluting buffer. Tris HCl 10 m*M*, pH 8, autoclaved solution.

Patch intracellular solution. 140 m*M* CsCl, 3 m*M* $MgCl_2$, 5 m*M* EGTA, 10 m*M* HEPES (pH 7.2). 80 ml of this solution is prepared as follows. CsCl, $MgCl_2$, and HEPES are dissolved in 60 ml H_2O and 2 ml of the EGTA (200 m*M* dissolved in 0.1 *N* KOH) stock solution is added. The pH is adjusted to 7.2 with KOH. The volume is then adjusted to 80 ml with H_2O and the solution is filtered (to remove any particles which may preclude gigaohm sealing) and autoclaved. 1-ml aliquots are then stored at –80°C until use. It is possible to replace CsCl with 140 m*M* KCl or potassium gluconate. In addition, neuronal tracers such as biocytin may be added in the patch intracellular solution to achieve morphological analysis. The composition of this intracellular solution may be changed. For example, the $MgCl_2$ concentration can be decreased and $MgCl_2$ can be added afterward to the RT reaction if necessary, to reach a final concentration of 2 m*M*. In theory, any patch pipette solution can be used, provided it does not contain inhibitors of the RT or PCR reaction. Because the RT reaction takes place after the recording, the final concentration of the reagents can be adjusted accordingly.

20× DTT (Dithiothreitol). A 1 *M* solution is prepared in water and filtered (Millex). The working solution (20×) is 0.2 *M* in H_2O stored as 50-µl aliquots in 1.5-ml screw-cap tubes under nitrogen at –80°C.

5× RT mix. Hexamer random primers (Boehringer Mannheim). Dissolved in Tris (10 m*M*, pH 8) at 5 m*M*. Deoxyribonucleotides -dNTPs- (Pharmacia) are each supplied as a 100-m*M* solution. A working RT mix solution (5×) of random primers and dNTPs is prepared in Tris (10 m*M*, pH 8) with random primers at 25 µ*M* and dNTPs at 2.5 m*M* each. This working mix is stored as 30-µl aliquots in 500-µl tubes at –20°C.

RNAsin and reverse transcriptase (RTase, Superscript II from Gibco BRL) are both stored at –80°C, each as 5-µl aliquots in 1.5-ml screw-cap tubes under nitrogen.

PCR buffers. The 10× *Taq* buffer supplied by Perkin-Elmer or Qiagen with 15 m*M* $MgCl_2$ included was found suitable for all PCR reactions.

100× dNTP PCR solution. This solution contains the four dNTP (dATP, dCTP, dGTP, and dTTP), each at 5 m*M* in 10 m*M* Tris, pH 8. This solution is used for the re-amplification of the products from the first PCR or in pilot experiments.

100× PCR oligonucleotide primer. The stock is kept at –20°C, undiluted. The working solution (100×) is diluted at 10 pmole/µl in Tris (10 m*M*, pH 8). This 100× solution is stored in 50-µl aliquots at –20°C.

200× PCR oligonucleotide primer and multiplex primer premix. When multiplex PCR involves a large number of primer pairs (e.g., 10 pairs), we prepare a 200× working solution for each separate primer as described above (20 pmole/µl) and mix equal volumes of each primer to prepare a premix: for 10 primer pairs (20 primers), we put 10 µl of this premix for each reaction tube (10 pmole of each primer).

PCR primer design. The choice of oligonucleotides is very important. In most instances, failure in pilot experiments could be related to a poor design of a primer pair (with stable hairpin or primer dimer combination). For multiplex PCR, the melting point of all primers must be between 55° and 60°C. We select primers generating cDNA amplified fragments ranging from about 200 to 600 bp. To avoid genomic DNA amplification from the nucleus, oligonucleotide primers are designed to amplify DNA sequences located on separate exons. If the gene structure is unknown, we proceed to amplifications of 200-ng genomic DNA and compare it with the cDNA amplification product.

5.2.2 Harvesting Procedure

The very first step of the single-cell RT-PCR protocol starts with the electrophysiological recording of the neuron. First, the cell is visually identified and chosen according to morphological criteria. The patch pipette filled with 8 µl intracellular recording solution approaches the neuron with a positive pressure in order to avoid contamination by the surrounding tissue of cells. Once the pipette reaches the neuron, a gigaohm seal is performed which ensures a very tight contact between the cell membrane and the pipette, avoiding the risk of contamination (cell attached configuration). Then, the patch membrane under the pipette is broken by applying a brief negative pressure, thus establishing physical and electrical continuity between the cytoplasm and the pipette solution. At the end of the recording, the cell content is aspirated into the recording pipette under electrophysiological and visual controls. The harvesting procedure is stopped if the gigaohm seal is lost. After cytoplasm collection, the pipette is gently withdrawn from the neuron to favor the closure of the cell membrane. Ideally, an outside-out configuration is then achieved, which allows the preservation of the pipette's content.

5.2.3 General Protocol for the RT Reaction

During the course of the electrophysiological experiments, the aliquots of intracellular solution, 5× RT mix, and 20× DTT are kept on ice, and RNAsin and RTase are kept at –20°C. During the recording, 2 µl of 5× RT mix and 0.5 µl of 20× DTT are pipetted into the RT-PCR tube. After recording and aspirating the cell, the pipette's contents are expelled (usually 6.5 µl) into this tube. We then add 0.5 µl RNAsin and 0.5 µl RTase, and the tube is flicked and briefly centrifuged. The final volume should then be approximately 10 µl, with final concentrations of 0.5 m*M* each dNTP, 5 µ*M* random primers, 10 m*M* DTT, 20 units of RNAsin, and 100 units of RTase. The capped tube is then stored overnight at 37°C. After this incubation, the tube is kept at –80°C until the PCR reaction.

5.2.4 General Protocol for the First PCR

Hot start option. Hot start always gives better results and is routinely used in our laboratory. In this option, primers should be added in the preheated RT-PCR tubes because the formation of primer-dimers is mainly due to the presence of RTase in the 10-µl RT reaction. To avoid this, we prepare two different solutions. Solution 1: volume is 70 µl × number of cell tubes. It contains per cell tube: 10 µl 10× PCR buffer, 0.5 µl *Taq* polymerase (2.5 units), and water up to 70 µl. Solution 2: volume is 20 µl × number of cell tubes. It contains per cell tube: 1 µl of the 100× sense primer solution, 1 µl of the 100× antisense primer solution (in case of multiplex 10 pmole of each primer), and water to 20 µl. 70 µl of solution 1 is added to the 10 µl RT reaction in each cell tube and overlaid with two drops of mineral oil. The tubes are then placed in the PCR machine preheated to 94°C, and after 1 min, 20 µl of solution 2 is added to each tube on top of the oil. After 1 min, the PCR program is started. The final volume in the PCR tube is 100 µl with final concentrations of

50 μM each dNTP (from the RT reaction), 10 pmole/100 μl of each of the primers, and 2.5 units/100 μl of *Taq* polymerase.

For PCR with only one primer pair, we perform 40 PCR cycles and the amplification products can be analyzed by gel electrophoresis with ethidium bromide staining. For multiplex PCR, 20 PCR cycles are performed, always followed by a 35-cycle PCR.

5.2.5 General Protocol for the Second PCR

For multiplex PCR (up to 40 primer pairs), 0.5 to 2 μl of the first PCR reaction without purification step is used as a template for the second PCR. Each of the cDNA species co-amplified during the first PCR will be amplified independently by the second PCR. To the 2 μl of the first PCR reaction, 78 μl are added, containing per tube 10 μl 10× *Taq* buffer, 0.5 μl *Taq*, 1 μl of the 100× dNTP solution and water. Two drops of oil are added. Then, the tubes are placed in the PCR machine preheated to 94°C (hot start protocol), and after 1 min, 20 μl containing 1 μl of the 100× sense primer solution, 1 μl of the 100× antisense primer solution, and water are added to each tube on top of the oil. After 1 min, the PCR program is started; 35 cycles are performed. The products of these second PCR amplifications are then analyzed by agarose gel electrophoresis stained with ethidium bromide.

5.2.6 Multiplex One-Step Amplification for Hybridization to the Neurochips

As in the two-round PCR amplification protocol, hot start is performed with the oligonucleotides mixture (solution 2 contains 5 μl with all the primers at 10 picomoles each and 15 μl water). Solution 1 contains 10 μl 10× buffer (Qiagen buffer containing KCl and $(NH_4)_2SO_4$ provides a wide temperature range for specific annealing of primers and a greater tolerance to variable Mg^{2+} concentration); 8 μl $MgCl_2$ (25 m*M*); 8 μl dATP and dGTP (2.5 m*M*); 7 μl dTTP and dCTP (2.5 m*M*); and 2.5 μl Cy3-dCTP and Cy3-dUTP (Amershan Pharmacia Biotech at 1 m*M*) with 2 μl of AdvanTaq from Clontech to give 80 μl total. After 45 cycles of amplification, PCR products are purified on Qiaquick columns (Qiagen), eluted in 30 μl, and 15 μl is used for hybridizing to the neurochips.

5.3 DESCRIPTION OF THE DNA MICROARRAYS: THE NEUROCHIPS

We have developed DNA chips containing 94 PCR fragments of 200 to 600 bp, corresponding to 94 rat genes important in neurotransmission. These neurochips include precursors for neuropeptides, enzymes involved in the synthesis of neurotransmitters, calcium binding proteins, and subunits of receptors for neurotransmitters: glutamate, GABA, acetylcholine, serotonine, dopamine, and adrenaline (Table 5.1). PCR fragments usually correspond to open reading frames and the sequence spans several exons when possible; the most specific gene sequence is

TABLE 5.1
List of Genes Spotted on Neurochips

Calcium binding proteins (3)	Calbindin, parvalbumin, calretinin
Enzymes synthesis (8)	GAD65, GAD67, CAT, TH, TpH, NOS-1, NOS-2, NOS-3
Peptides (6)	NPY, VIP, somatostatine, CCK, endorphin, enkephalin
AMPA receptors (4)	GluR1, GluR2, GluR3, GluR4
Kainate receptors (5)	KA1, KA2, GluR5, GluR6, GluR7
NMDA receptors (4)	NR2A, NR2B, NR2C, NR2D
Metabotropic receptors (8)	mGluR1, mGluR2, mGluR3, mGluR4, mGluR5, mGluR6, mGluR7, mGluR8
GABA receptors (13)	α1, α2, α3, α4, α5, α6, β1, β2, β3, γ1, γ2, γ3, δ1
Nicotinic receptors (10)	α2, α3, α4, α5, α6, α7, α9, β2, β3, β4
Muscarinic receptors (5)	m1, m2, m3, m4, m5
Dopamine receptors (5)	D1, D2, D3, D4, D5
5-HT receptors (13)	5-HT1A, 5-HT1B, 5-HT1D, 5-HT1E/F, 5-HT2A, 5-HT2B, 5-HT2C, 5-HT3, 5-HT4, 5-HT5a, 5-HT5b, 5-HT6, 5-HT7
Adrenaline receptors (9)	α1a, α1b, α1c, α2b, α2c, α2d, β1, β2, β3
Controls (1)	Somatostatine intron
Normalizing genes (5)	G3PDH, tubulin, actin, ubiquitin, HPRT

chosen. These rat neurochips have been shown to work with human and mice mRNA samples. The list of gene fragments and PCR primers can be obtained from the Web site: http://www.espci.biologie.fr. The T_m of the primers is between 55 and 65°C.

PCR fragments (DNA targets) were first amplified from 15-day-old rat brain cDNAs and then reamplified and purified on Qiaquick columns (Qiagen). The final concentration of DNA targets before spotting was 100 ng/µl in 3×SSC. All the DNA targets were reamplified similarly, indicating that the 94 PCR reactions worked efficiently.

For all the genes of the same family, the percentage of identity has been calculated. Table 5.2 gives the results for the AMPA/kainate family. The asymmetry of the table is due to the variable positions of DNA targets along the genes. We have hybridized a 2540-bp fluorescent PCR probe corresponding to the GluR3 receptor subunit to microarrays containing all the AMPA and kainate receptor subunits listed in Table 5.2 at various stringency conditions (hybridization buffer and temperature, washing buffer and temperature). We showed that cross-hybridizations occur with GluR2 and GluR4 subunits and that they are at the lowest when using SSC/SDS buffer (3.5× SSC, 0.3% SDS) at 60°C instead of formamide (50%) at 42°C and when washing the slides at 65°C in 0.05×SSC. No cross-hybridization was detected with GluR1 despite a 73% homology with GluR3. The difference in cross-hybridization observed between GluR1 and GluR2/GluR4 is explained by the repartition of homology along the DNA sequence. A 70% homology corresponds to either a 70% homology all along the DNA sequence, or to a 100% homology on 70% of the sequence. In conclusion, for minimizing cross-hybridization, the percentage of identity should be less than 70% and homogeneously distributed.

TABLE 5.2
Percentage of Homology Between DNA Targets (horizontal column) and the Corresponding DNA Fragments in the Genes of the Same Family (vertical column)

	Glur1	Glur2	Glur3	Glur4	Glur5	Glur6	Glur7	Ka1	Ka2
Glur1	100%	54%	53%	53%	45%	42%	41%	38%	38%
Glur2	60%	100%	60%	62%	39%	37%	37%	34%	33%
Glur3	73%	74%	100%	75%	56%	61%	55%	55%	53%
Glur4	77%	76%	81%	100%	45%	47%	46%	45%	46%
Glur5	61%	60%	60%	60%	100%	81%	80%	65%	67%
Glur6	36%	38%	39%	41%	65%	100%	53%	41%	40%
Glur7	39%	40%	39%	36%	70%	72%	100%	40%	41%
Ka1	36%	34%	35%	35%	37%	38%	37%	100%	61%
Ka2	36%	36%	37%	34%	41%	40%	43%	69%	100%

5.4 ANALYSIS OF GENE EXPRESSION IN BRAIN EXTRACTS WITH THE NEUROCHIPS

Differential expression can be measured with the neurochips using cDNAs from tissue samples labeled with a fluorochrome (Cy5) during reverse transcription of mRNAs, hybridized to the arrays together with cDNAs from a control tissue labeled with another fluorochrome (Cy3). With this type of experiment, it is possible to detect a 1.5-fold variation in the level of expression of any gene that is sufficiently transcribed. Just 1 μg of polyA$^+$ is necessary to obtain a strong fluorescence signal. For cDNA labeling and microarray hybridization, we followed the protocols from the P. Brown Laboratory, available at http://www.cgm.stanford.edu/pbrown.

In collaboration with the Neuroscience Research Center of Merck Sharp & Dohme (B. Le Bourdellès, T. Rosahl, and P. Whiting), we have studied the expression pattern of the 94 genes contained on the neurochips in mice in which the β2 subunit of the GABA-A receptor was knocked out (KO). The most prominent GABA-A receptor combination in mammalian brain is α1/β2/γ2. In the β2 KO adult mice, approximately 50% of the H^{3}Flumazenil binding sites are lost. However, the β2 homozygotes appear phenotypically normal, with slightly increased activity. A 50% loss of inhibitory receptors could be compensated by an under- or over-expression of other receptors, resulting in the absence of a severe phenotype in the animals. We used the neurochips to study a putative compensatory mechanism. Experiments were performed on two adult animals (at least 7 months old). For the normalization purposes, we added actin, tubulin, HPRT, ubiquitin, and G3PDH DNA targets onto the neurochips. G3PDH was found to be the best normalizing gene in rat brain despite its high level of expression compared to most of the genes present on the neurochips.

Figure 5.1A shows an example of differential gene expression in KO mice (red) vs. control mice (green) analyzed by neurochips and presented as a pie chart. Figure 5.1B demonstrates the expression of GABA-A receptor subunits only in these animals. Pies have been calculated using a program written by P. Marc and C. Jacq from the Ecole Normale Superieure in Paris (http://www.biologie.ens.fr/microarrays.html). The size of the pies is proportional to the sum of the fluorescence intensities in KO and control mice. Figure 5.1B shows that the γ2 subunit is downregulated, together with other subunits that are clustered on the same chromosome in the order β2 α6-α1γ2.[11] The presence of the neomycin cassette in the β2 subunit knockout construct allows only partial suppression of this gene at the level of transcription, but resulted in an inhibition of translation of mRNAs into proteins. In addition, we have detected a reproducible downregulation of calcium binding proteins (calbindin and calretinin) with a ratio higher than 1.5. Glutamate receptor subunits GluR4, GluR6, and mGluR6, muscarinic receptor 2, and the 5HT5a serotonin receptor could also be downregulated. We have compared the ratios obtained

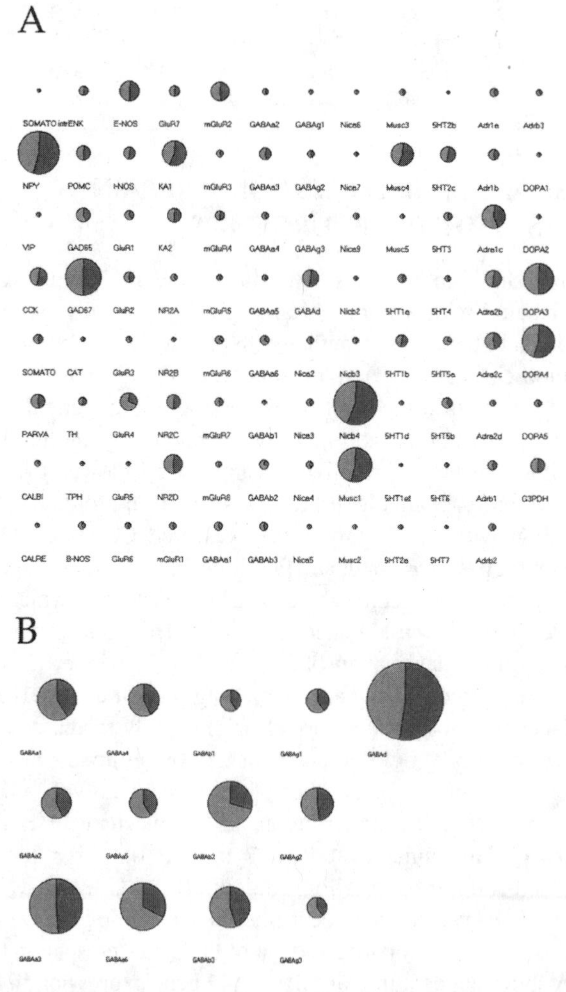

FIGURE 5.1 Differential gene expression profile of brain tissue of knockout mice for the GABA-β2 subunit analyzed by the neurochip. (A) 94 genes present on the neurochips. (B) Subset of GABA-A receptor subunits.

TABLE 5.3
Comparison of the Ratios of Expression of Various GABA-A Receptor Subunits Using Neurochips and Northern Blots in Control vs. β2 Knockout Mice

GABA-A Receptor Subunit	Neurochips	Northern Blots
α1	0.69–0.66	0.65
α6	0.47–0.5	0.22–0.21
β2	0.39–0.21	ND
γ2	0.91–0.71	1.13

for α1, α6, and γ2 GABA-A receptor subunits using neurochips and northern blots. Table 5.3 shows the values obtained from the same animals. Because the ratios are comparable or lower on northern blots, we conclude that the results obtained on neurochips were confirmed using northern blots.

5.5 ANALYSIS OF GENE EXPRESSION IN SINGLE CELLS WITH THE NEUROCHIPS

In 1992, Lambolez and Audinat[4] developed the technology that allows one to correlate electrophysiological properties of single neurons with the gene expression profile in single neurons — single-cell RT/PCR. Application of RT/multiplex PCR allows analysis of 30 genes in a single cell.[10] After patch-clamp recording, the cytoplasm of the cell is harvested through the recording pipette and the mRNA content is reverse-transcribed into cDNAs. cDNAs are amplified in two rounds of PCR; the first round consists of 20 cycles multiplex PCR and amplifies all 30 genes at the same time using degenerate primers. In the second round, primers specific for each gene consist of 30 different PCRs of 35 cycles, each one specific for one of the 30 genes. PCR products of the second round of PCR have to be analyzed by agarose gel electrophoresis (Figure 5.2).

For studying a larger number of genes at the single-cell level, we are now using the neurochips. To test the multiplex PCR for amplifying 94 genes in the same reaction, we performed some hybridizations with mRNAs or cDNAs from the same 15-day-old rat brain. Experiments were done at least twice on two different sets of microarrays obtained from different batches of DNA targets spotted with the two spotters described above. The 94 DNA targets were amplified in 94 separate PCR reactions from cDNA quantities corresponding to a one-cell content (the equivalent of about 1 pg mRNA). PCR products were analyzed on agarose gels. No detectable band was given the value 1. Values 2, 3, 4, and 5 were attributed to genes giving, respectively, low-, medium-, high-, and very high-intensity bands on agarose gel after staining with ethidium bromide. This experiment, called "agarose RT-PCR," is depicted in Figures 5.3 and 5.4 (axis 2).

In another experiment, the 94 DNA targets were amplified in 94 separate PCR reactions in which Cy5-dCTP was incorporated. Fluorescent DNA targets were

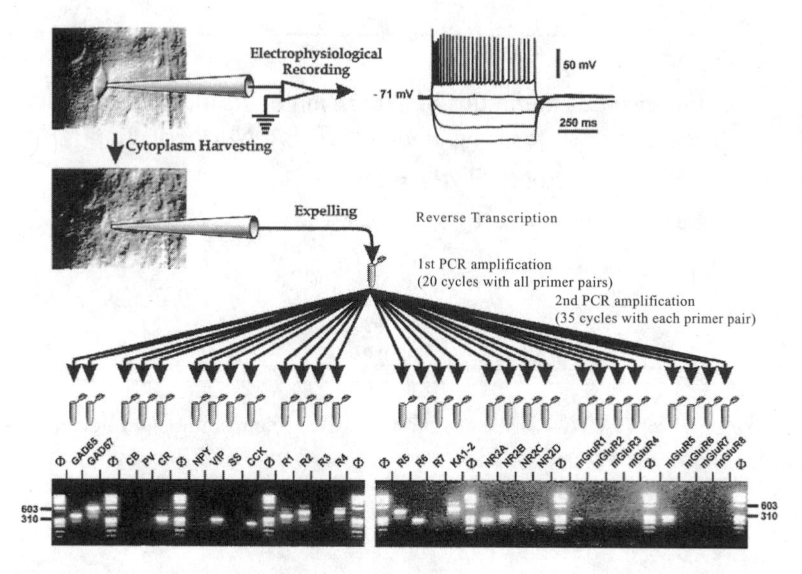

FIGURE 5.2 Scheme of the reverse-transcriptase multiplex PCR utilized for analysis of gene expression of single cells.

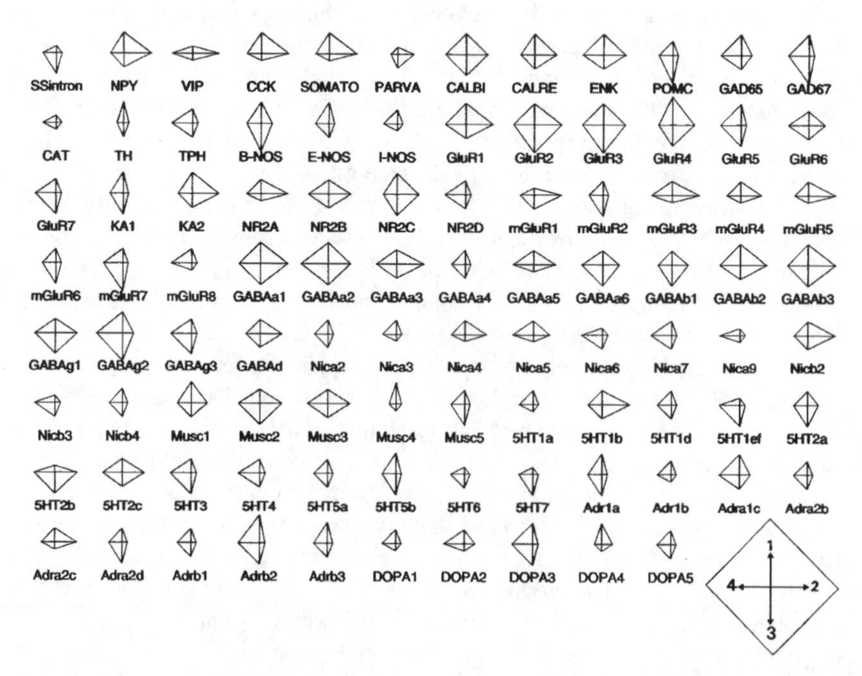

FIGURE 5.3 Star plot for analyzing the hybridization signals from neurochips. Each axis represents different characteristics for detection of the gene expression profile. Axis 1, fluorescent cDNAs; axis 2, agarose RT-PCR; axis 3, size of the DNA targets; axis 4, fluorescent PCR probe.

A

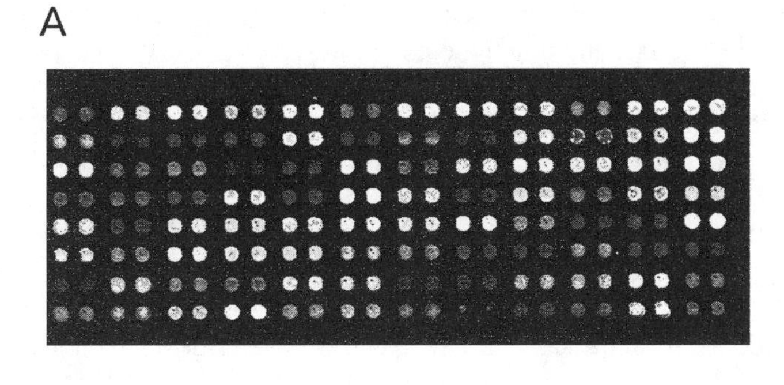

B

FIGURE 5.4 Analysis of gene expression in single cell using neurochip. (A) Hybridization to neurochips of the fluorescently labeled multiplex PCR products obtained from cDNA quantities corresponding to a one-cell content. (B) Star plot analysis of multiplex PCR products (94 genes) amplified in one tube. Axis 1, multiplex PCR from DNA targets; axis 2, agarose analysis of RT-PCR; axis 3, second round of multiplex PCR from cDNA obtained in the first round of amplification. (See Color Figure 5.4 following page 82.)

mixed, purified on a Qiaquick column, and hybridized to the neurochips in 3.5× SSC/0.3% SDS at 60°C in a humidity chamber overnight. Microarrays were washed at room temperature in 0.05× SSC and the fluorescence signal was analyzed on the Scanarray 3000 (General Scanning). Fluorescence signals up to 3000 per spot were given the value 1, from 3000 to 10,000 the value 2, from 10,000 to 30,000 the value 3, from 30,000 to 50,000 the value 4, and from 50,000 to 65,534 the value 5. This experiment is called "fluorescent PCR probe" in Figure 5.3 (axis 4). We then amplified the 94 genes in a multiplex PCR in one tube. Each DNA target (10 fg) was mixed and amplified in one 45-cycle multiplex PCR experiment containing the 94 primer pairs (10 pmoles each), 3.5 m*M* $MgCl_2$, 200 μ*M* dATP and dGTP, 160 μ*M* dTTP and dCTP, and 40 μ*M* Cy3-dCTP and Cy3-dUTP, with the Advantaq from Clontech in 100 μl in the Qiagen *Taq* buffer containing KCl and $(NH_4)_2SO_4$ providing a wider temperature window for specific annealing of primers and a greater tolerance to variable Mg^{2+} concentration. Fluorescent PCR fragments were purified on a Qiaquick column (Qiagen, Valencia, CA) and eluted in 30 µl TE. A 10-µl aliquot was adjusted to 3.5× SSC and 0.3% SDS (final concentrations), and then hybridized to the microarrays analyzed as described above. This experiment is called "multiplex PCR from DNA targets" in Figure 5.4 (axis 1).

The same multiplex PCR was performed starting with cDNA quantitites corresponding to a one-cell content (the equivalent of about 1 pg mRNA). This experiment is called "multiplex PCR from cDNA" in Figure 5.4 (axis 3). Finally, fluorescent cDNAs produced from the same mRNA batch were hybridized to the neurochips as described above. This last experiment is called "fluorescent cDNAs" (Figure 5.3, axis 1).

All these experiments were analyzed using the R software available under the GNU General Public License at: www.gnu.org. In Figure 5.3 the size of the DNA target was added (1 for 0 to 200 bp, 2 for 200 to 300 bp, 3 for 300 to 400 bp, 4 for 400 to 500 bp, and 5 for 500 to 600 bp) (see axis 3 in Figure 5.3).

Figure 5.3 shows that the hybridization signal of a mixture of fluorescent target DNAs ("fluorescent PCR probe") is not correlated to the length of the target DNA. For example, the DNA target corresponding to the VIP gene (287 bp) gives an intense fluorescent hybridization signal, while the DNA target corresponding to parvalbumin (PARVA, 388 bp) gives a low signal. From this first series of experiments, we can detect DNA targets that do not hybridize very well and choose another one in the corresponding gene sequence.

Figure 5.4(A) is an example of hybridization to the neurochips of fluorescent multiplex PCR products obtained from cDNA quantities corresponding to a one-cell content. Rainbow colors correspond with the fluorescence signal: from purple to white for increasing intensities. The microtiter plate-like organization of the genes in Figures 5.3 and 5.4(B) corresponds to the order of spotting in Figure 5.4(A) where genes were spotted in duplicates. Figures 5.4(A) and (B) show that 20 genes (underlined) out of 94 are not correctly amplified in the multiplex protocol ("multiplex PCR from DNA targets" experiment). When expressed in rat brain and detected in the "agarose RT-PCR" experiment, these 20 genes will not be detected in the multiplex experiments. Figure 5.4 also shows that some genes give a perfect equilateral triangle. Although a multiplex PCR for amplifying 94 genes in the same reaction will never be quantitative, we do observe that the expression levels of 12 genes are largely underestimated. One example is GluR1, for which the hybridization of fluorescent cDNAs or fluorescent PCR probe is correct (Figure 5.3) but the multiplex PCR is not very efficient (Figure 5.4).

Finally, for three genes, the expression levels deduced from hybridization of fluorescent cDNAs are higher than from agarose RT-PCR experiments (Figure 5.3): GluR7, GAD67, and Adrβ2. This discrepancy could be explained by a cross-hybridization of GAD67 with GAD65, of GluR7 with GluR5, and of Adrβ2 with another gene.

5.6 CONCLUSIONS

Similar to the development of the PCR or the recombinant DNA technology, microarrays have a large number of applications that will expand over time. In neurobiology, because of the very high cellular diversity of the brain, it is very important to develop techniques that will be applied to the study of gene expression at the single-cell level. Because the mRNA content of individual cells is much less complex than the mRNA content of a whole brain, we can predict that the multiplex PCR protocol can be used for amplifying 94 genes from individual cells in one reaction and hybridizing fluorescent PCR products to the neurochips. These experiments will provide yes or no answers but no quantitative data. An alternative is the protocol described for amplifying 94 genes from a single cell in two steps: the first step of multiplex PCR in one tube and the second step of 94 individual PCR in a microtiter plate. This protocol has been used for analyzing the expression profiles of Purkinje cells.

It is important when designing new microarrays to work out the problem of specificity among gene families. With RT/PCR, specificity relies on the sequence of the primer pairs. With DNA microarrays, specificity relies on the sequence of the hundreds of DNA targets. The same neurochips are being used to study differential gene expression in a variety of KO mice. The targeted genes include subunits of GABA-A, glutamate, dopamine and nicotinic receptors, and the calcium binding proteins calretinin, calbindin, and parvalbumin.

ACKNOWLEDGMENTS

The authors would like to thank Mike Eisen for very useful advice and Alex Papanastassiou for participating in the production of DNA targets. This work was supported by CNRS (Genome Project), ARC, and Curie Institute Funds.

REFERENCES

1. Schena, M., Shalon, D., Davis, R., and Brown, P., Quantitative monitoring of gene expression patterns with a complementary DNA microarray, *Science*, 270, 467–470, 1995.
2. Eisen, M. and Brown, P., DNA arrays for analysis of gene expression, *Methods Enzymol.*, 303, 179–205, 1999.
3. Eisen, M., Scanalyze at cgm.stanford.edu/pbrown.
4. Lambolez, B., Audinat, E., Bochet, P., Crepel, F., and Rossier, J., AMPA receptor subunits expressed by single Purkinje cells, *Neuron*, 9, 247–258, 1992.
5. Lambolez, B., Ropert, N., Perrais, D., Rossier, J., and Hestrin, S., Correlation between kinetics and RNA splicing of -amino-3-hydroxy-5-methylisoxazole-4-propionic acid receptors in neocortical neurons, *Proc. Natl. Acad. Sci. U.S.A.*, 93, 1797–1802, 1996.
6. Angulo, M.C., Lambolez, B., Audinat, E., Hestrin, S., and Rossier, J., Subunit composition, kinetic, and permeation properties of AMPA receptors in single neocortical nonpyramidal cells, *J. Neurosci.*, 17, 6685–6696, 1997.

7. Cauli, B., Audinat, E., Lambolez, B., Angulo, M.C., Ropert, N., Tsuzuki, K., Hestrin, S., and Rossier, J., Molecular and physiological diversity of cortical nonpyramidal cells, *J. Neurosci.*, 17, 3894–3906, 1997.
8. Porter, J.T., Cauli, B., Staiger, J.F., Lambolez, B., Rossier, J., and Audinat, E., Properties of bipolar VIPergic interneurons and their excitation by pyramidal neurons in the rat neocortex, *Eur. J. Neurosci.*, 10, 3617–3628, 1998.
9. Porter, J.T., Cauli, B., Tsuzuki, K., Lambolez, B., Rossier, J., and Audinat, E., Selective excitation of subtypes of neocortical interneurons by nicotinic receptors, *J. Neurosci.*, 19, 5228–5235, 1999.
10. Cauli, B., Porter, J.T., Tsuzuki, K., Lambolez, B., Rossier, J., Quenet, B., and Audinat, E., Classification of fusiform neocortical interneurons based on unsupervised clustering, *Proc. Natl. Acad. Sci. U.S.A.*, 97, 6144–6149, 2000.
11. Kostrzewa, M., Köhler, A., Eppelt, K., Hellam, L., Fairweather, N.D., Levy, E.R., Monaco, A.P., and Müller, U., Assignment of genes encoding GABA-A receptors subunits $\alpha 1$, $\alpha 6$, $\beta 2$ and $\gamma 2$ to a YAC contig of 5q33, *Eur. J. Hum. Genet.*, 4, 199–204, 1996.

6 High-Dimensional Visualization Support for Data Mining Gene Expression Data

Georges Grinstein, C. Bret Jessee, Patrick Hoffman, Phil O'Neil, and Alexander Gee

CONTENTS

0-8493-2285-5/02/$0.00+$1.50

6.1 INTRODUCTION

6.1.1 BACKGROUND

Numerous examples abound on the use of specific algorithms or techniques to analyze the results of specific array experimental data and several provide comparative studies.[1–27]

Given such a dataset, which algorithm should we select in mining gene expression data? Are we looking to cluster? To classify? To identify associations or other relationships? Are we looking to build a predictor? To analyze time series? Are we verifying a hypothesis? All these questions are relevant and each is repeated numerous times in every area of bioinformatics and cheminformatics. The large number of available techniques to choose from does not simplify the problem.

In some fashion, any approach will generally work — to some extent. A sufficient data mining technique in the hands of a sufficient miner with a sufficient amount of time will yield reasonably good results. Sufficient is clearly vague and depends on the task at hand. It could be quantified. However, the point we are trying to make is that most common structures in data are fairly easy to discover. It is their separation — their extent — we seek to improve. It is also the uncommon structures and those structures that are eclipsed by the common ones (the nuggets) that we seek.

The most important goals are reducing the time to the discoveries and improving the precision, validity, and reliability of the results.

The cost-savings that can be incurred by eliminating errors in certain steps within the discovery process can be dramatic (well over $1 million per day!). Thus, reaching 95% and up in confidence, validity, or whatever metric is used, may be necessary as well in many of the discovery problems.

Thus, if reducing time or increasing the quality of the results to a maximal extent is desired, the process, tools, and the miner's expertise are important and critical: a rich set of tools (much better than sufficient), expertise (again better than sufficient), and a comprehensive data exploration process are necessary.

6.1.2 IMPROVING THE PROCESS

The process of discovery is still often incomplete. Most data miners frequently arrive at dramatically different results with different algorithms. Witness the various different collections of genes identified as cancer classifiers derived from multiple analyses of the Golub and Slonim et al. dataset.[13,28] Thus, no single answer is comprehensive and all seem to provide some element of truth. The fundamental problem is that we are attempting to discover the logical connections between multiple protein functions and the state(s) of a biological system. This is an attempt to model a system through a small set of incomplete views divulging indirect information about structure and function. This indirect information is the measured variation of gene expression in organisms under a variety of, and across, experimental natural or pathological conditions.

The connection between gene expression patterns and gene function has now been widely demonstrated through the application of cluster analyses to gene expression datasets. As such, these results are all valid within some constraints. This again emphasizes the need for a method of discovery that encompasses a variety of approaches, all involving the user in the discovery and decision process. The user must inject knowledge about the domain to resolve the ambiguities in the discovery process.

The datasets we are dealing with are becoming massive (gigabytes and terabytes), with thousands of variables and millions of records. These have devastating computational ramifications. It is tempting to attack the problem with purely analytical techniques supplemented with distributed or high-performance computing,[29–33] arguing that the data is just too massive for human comprehension, let alone for human analysis. But this increasing separation of the user in the computational process as well as the discovery process makes it difficult if not impossible to resolve the above ambiguities and, even more so, harness human intuition. The user is simply more and more left out of the loop.

There is an additional problem resulting from sampling statistics. When testing variables for significance, for example, in discriminating two classes, it is important to be aware of the false positives. For example, if a test was performed on 6800 variables (for the example data set described below) with a significance level of 0.01, one would expect that about 68 would pass the test if the data was random. Even at the 0.001 level, one would expect to find 5 or 10 passing by chance. This means that not only is statistical significance important, but also that the biological meaning must be evaluated in all results.

Several systems do attempt to keep the user in the loop, but in an extremely demanding manner; and it is a particular kind of user being involved, one who is very knowledgeable in the process, the domain, and even more so the algorithms. This works out well for very specific tasks. Training on such systems can be developed and thus expertise transferred. However, again, this is for very specific applications.

Consider, for example, the NCI tool for producing a cluster image map.[34,35] When not using the defaults, the user can select a centering technique (two choices), whether to randomize (two choices), the cluster method (three choices), and the cluster distance metric (six choices for each axis). This yields a total of 432 different computations, each with additional variable parameters (the proportion to remove and which subset of the cells to use), and each with a large number of possible displays (windows, trees, colors, scales, and data ranges). This is staggering and clearly shows the scale of the problem that confronts us.

This impediment is due to the lack of common usage and understanding of very high-dimensional visualization. This approach is now a necessity to grasp the meaning inherent in massive datasets which is hidden from human perception, recognition, and understanding by high dimensionality.

This chapter describes the process of discovery and focuses on that high-dimensional visualization support.

Thus, in summary, it can be argued that most algorithms or techniques will work sufficiently well in producing good analytical results with a knowledgeable data miner at the helm. These results would reach levels of confidence and reliability approaching the 85 to 90% range. It can also be argued that to dramatically increase the levels of confidence to well above the 95% threshold, that level justifying the hundreds of millions of dollars and years of effort needed to reach the market with new drugs and diagnostics, will require interactive tools capable of tuning the various analytic tools and the visualization support to tune them.

6.2 THE DATA MINING TASKS

The key data mining tasks are summarization or characterization, association, classification or prediction, and clustering, although there is a tremendous amount of overlap. Because these topics are well known and not the focus of this chapter, we provide a simple description with numerous references.

6.2.1 Summarization and Characterization

Summarization techniques attempt to provide overviews of the structure of the data and are often the focus in the processing and massaging of the data for downstream, more detailed mining. Outlier or deviation detection, for example, fits within this task (characterization) whereby one wants to identify and explore the reasons for the deviations.[36–44,200] Many of the techniques within this task can be found in the clustering literature as well (Section 6.2.4).

6.2.2 Association

Link analyses techniques attempt to identify the connections or relationships between records in the data. These are often called association rules or, more simply, associations.[14,45–56] Once association rules have been discovered, they are mined using any of the other mining techniques. See also the clustering literature (Section 6.2.4).

6.2.3 Prediction or Classification

Predictive modeling techniques or classifiers attempt to build a model of the data in order to classify additional data. This is often called classification. There are numerous automatic and user-driven techniques and, again, many are found in the clustering literature (Section 6.2.4).[57–76]

6.2.4 Clustering

Clustering determines a domain-dependent natural and useful partitioning of a data set into a number of clusters.[77–86] It is important to state that clustering has its origins in statistics and much computational and theoretical work is still based there.[87–126] Many new results are appearing that provide a firmer foundation for the various results of clustering. For example, once a cluster structure is discovered, its validity (or statistical significance) needs to be computed. One must ask: could this be a random finding?

Two other, related disciplines have developed very special clustering approaches: (1) databases, especially as applied to very large and massive databases,[84,85,127–138] and (2) visualization.[139–141] There are also specialized clustering techniques such as hierarchical[142-146] and self-organizing maps (SOMs).[147–150] In hierarchical clustering, a tree-like structure (or forest, a collection of trees) is developed. This partitions the data in a parent/child relationship (or dependency) based on some measure of proximity. In an SOM, the data is segmented or partitioned into subsets consisting of similar records with a similarity based on a domain-dependent metric.

Many of the techniques can be applied across tasks. There are model-based approaches,[25,151–153] statistical techniques,[25,106,151,154–161] and artificial intelligence techniques, many falling under the rubric of machine learning.[12,18,162–172]

6.2.5 Data Mining Issues

Different mining techniques, different metrics, and different organizations of the data can yield different results. Clusters and associations may be interesting or not, as well as valid or invalid. Missing value handling is critical with data resulting from biological experiments. Associations may be trivial. All this implies that human participation as well as statistical validation are critical. One way to provide that participation is to integrate visualization into the process and increase the use of statistics. The importance of integrating visualization with data mining was identified early by the statisticians.[173–185]

Harnessing the human data explorer and analysis through integrated visualization in the knowledge discovery process has been discussed extensively by Grinstein in several papers.[186–196]

6.3 AnVil APPROACH TO DATA DISCOVERY

Our goal is speedy and continual customer-driven (or other) improvement and refinement of the discoveries. This consists of several steps and pushes the discoveries to the limit realizable.

The process begins with summarization and characterization activities. Normalization and missing values are first handled and evaluated in the context of visual and statistical overviews of the complete data. There are several normalizations, each with their advantages. Normalization can be local on a per-row or per-column basis or global; it may be nonlinear; and it may be based on distributions. Missing values can be handled in numerous ways, from not using any records with missing values (which can in some cases reduce the dataset to nothing) to complex modeling of the non-missing value data. We typically first impute with an extreme value to clearly identify the missing data[197,198] and then impute values generated from the multidimensional modeling of the data.

Datasets can be further preprocessed by transformation, thresholding, and filtering, although appropriate methods remain equivocal. The application example below utilizes a published dataset without preprocessing to emphasize both the strength of visual analysis of unreduced data of high dimensionality as well as to avoid outcomes biased by different preprocessing approaches.

Full views of the data are then presented to the user for evaluation and tuning. Outliers are then further explored. Statistical metadata is then computed and also presented. This produces clusters and additional outliers to be monitored throughout the process.

We then proceed to mine the data in an attempt to cluster, classify, or develop association rules. We use visual techniques as an aid to the data explorer. These can be considered visual buffers for use as reference or base points. Users can easily build mental models of their data with these visualizations.

The statistical, machine learning, and visualization tools selected depend on the tasks. Figure 6.4 describes the process when building a predictor.

6.4 APPLICATION EXAMPLE

We describe a data mining approach focused on massive datasets produced by high-throughput microarray studies. Using high-dimensional visualization in support of data mining and analysis, we describe results that both inform on potential diagnostic and drug target gene sets as well as report on the quality of sample data. The approach allows for an iterative exploration of data and results to assess sample quality impact on results.

6.4.1 Critical Assessment of Techniques for Microarray Data Analysis December 2000 Meeting

The Critical Assessment of Techniques for Microarray Data Analysis (CAMDA00) meeting (December 18 and 19, 2000) presented the Golub, Slonim et al.[13] dataset as a common starting point for comparison of microarray analytical methods.

As groups variably applied preprocessing (thresholding, filtering, etc.) and different gene selection algorithms in construction of predictor sets, common as well as unique genes were identified in each predictor set. Few authors have presented the genes composing their derived class predictor sets; and of the dozens of groups that have published analyses of the Golub and Slonim data set, including the original authors, none have been able to present a gene predictor set diagnostic of clinical treatment outcome. We describe a 76-gene set that predicts the success or failure of chemotherapy treatment outcome for acute myeloid leukemia.

While a survival prognosis predictor has been generated for lymphoma,[79] a gene predictor set for prediction of the success or failure of chemotherapy treatment currently represents our unique contribution.

The standard in microarray analysis is to gather massive datasets and reduce them prior to clustering to find useful relationships, but this essentially discards the bulk of data that was gathered at great expense and cost of time. Our methods of analysis and visualization retain the full dataset to preserve all available information in making initial assessments of the data. Only after the full datasets are explored, missing values imputed, and outliers identified and tagged is the data dimensionally reduced for finer discrimination.

We first describe the two-class cancer dataset and Golub and Slonim et al.'s (hereafter called Golub and Slonim) approach, describe our approach, and detail the results.

6.4.2 Golub and Slonim Dataset

Golub and Slonim collected a group of 38 tissue samples from repositories of clinical samples taken from patients with acute lymphoblastic and acute myeloid leukemia (ALL and AML, respectively), and conducted DNA microarray analysis of gene expression profiles using the Affymetrix Hu6800 GeneChip. An independent test set consisted of 34 samples. Each sample consisted of a log-normalized expression value of the genes.

In many of the classifiers and predictors that will be produced, there is the issue of false positives, which are genes that would have been deleted from consideration by preprocessing. They might be "real" finds that were trimmed out or just noise. Given the poor quality of the Golub and Slonim dataset, it will be difficult to discern.

6.4.3 Golub and Slonim Approach

The authors presented their methodology (flowcharted in Figure 6.1) for building a cancer class predictor from known cases for subsequent use in classifying "unknown" samples as ALL and AML. They further presented a method for cancer class discovery *a priori*, without training via the use of known samples or classes. While perhaps unique at the time, this seminal publication has become the generic reference for class prediction and discovery from microarray datasets (Figure 6.1). A class predictor was constructed from data derived from this 38-sample training set and used to predict cancer classification on the independent test set of 34 samples.

Using correlation statistics and nearest-neighbor analysis of known classes and training set microarray data, Golub and Slonim arrived at a set of 50 genes to predict classification of patients in a test set (see Figure 6.2).

While one might have wished for genes always "on" in one cancer type and always "off" in the other, there appears to be no black-and-white answer in the real world of gene expression profiling. The upper left panel in Figure 6.3 depicts 25 genes selected as strongly expressed in ALL and weakly expressed in AML, with the lower left panel depicting 25 genes selected as strongly expressed in AML and weakly expressed in ALL. The discrimination favors strongly expressed genes as more weakly expressed genes fall below the S/N ratio. When applied to the test set, our approach is sufficient to accurately predict patient assignment to ALL or AML.

6.5 THE AnVil APPROACH

We depart from the now-traditional analysis methods exemplified by Golub and Slonim's analysis of microarray data by first taking a high-level overview of the dataset, examining statistical metadata and working in very high-dimensional space to assess the full dataset.

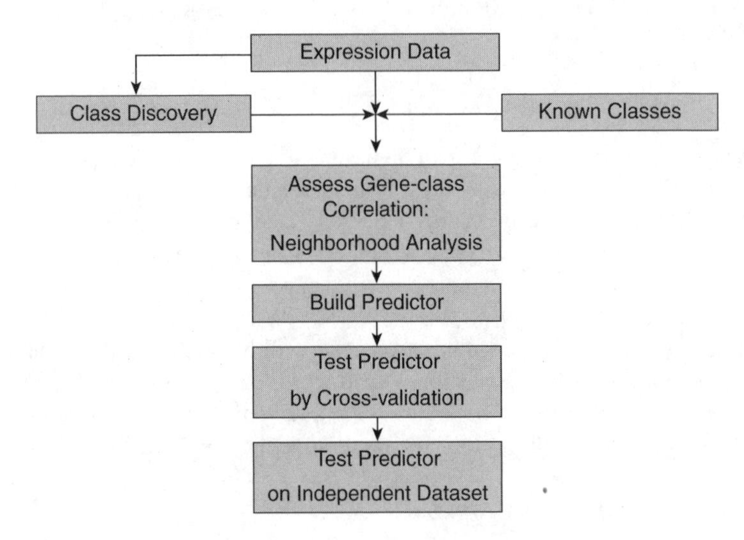

FIGURE 6.1 Golub and Slonim data discovery process.

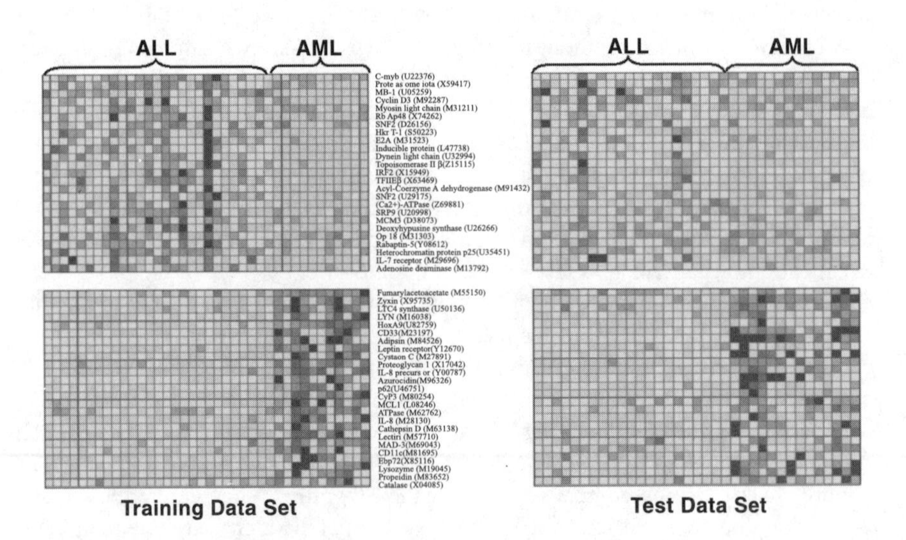

FIGURE 6.2 50 Set of 50 selected genes. (From Golub, T.R. et al., *Science*, 286, 531–537, 1999. With permission.)

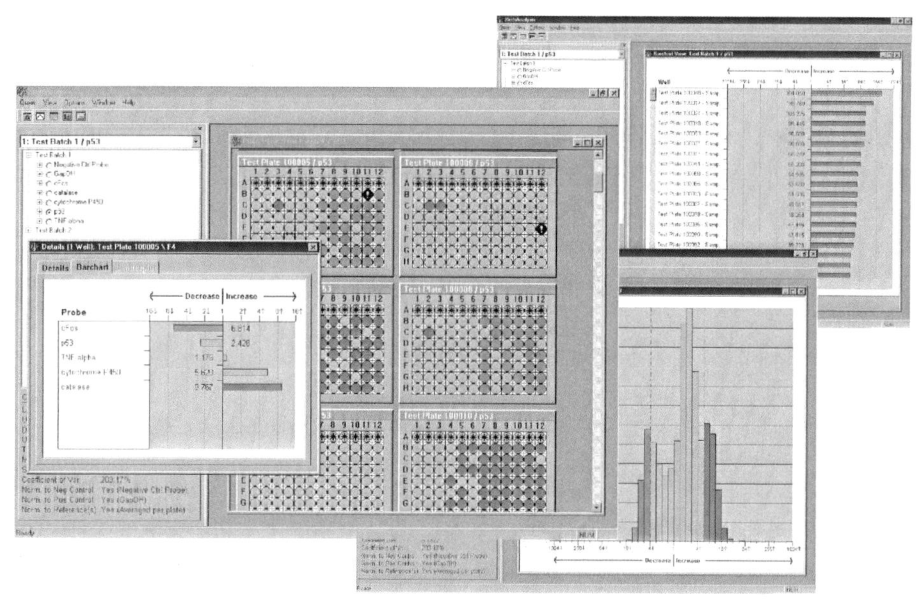

COLOR FIGURE 4.14 Instrument display screen showing results from one well.

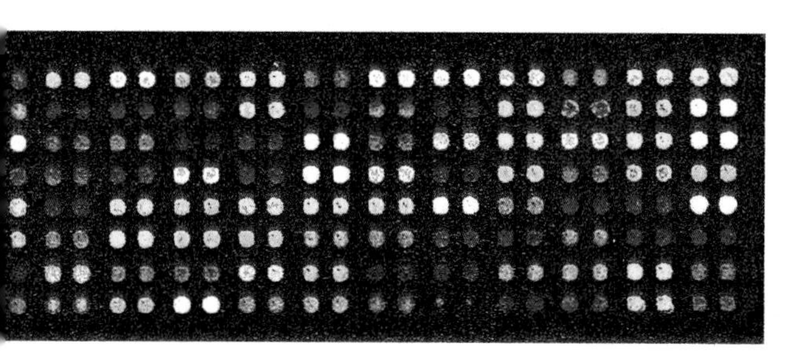

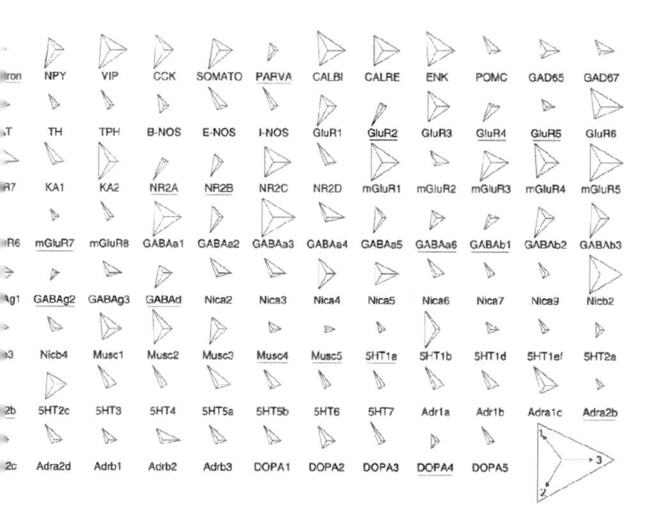

COLOR FIGURE 5.4 Analysis of gene expression in single cell using neurochip. (A) Hybridization to neurochips of the fluorescently labeled multiplex PCR products obtained from cDNA quantities corresponding to a one-cell content. (B) Star plot analysis of multiplex PCR products (94 genes) amplified in one tube. Axis 1, multiplex PCR from DNA targets; axis 2, agarose analysis of RT-PCR; axis 3, second round of multiplex PCR from cDNA obtained in the first round of amplification.

Pearson Correlation Coefficient
Gene Expression Levels

Statistical
Metadata

Absolute

Binned

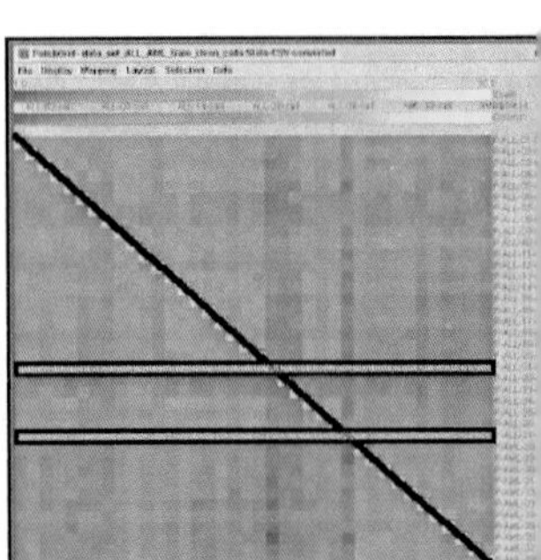

Patient 9, 17, 20, 21, and 27

COLOR FIGURE 6.8 Summary of metadata analysis results.

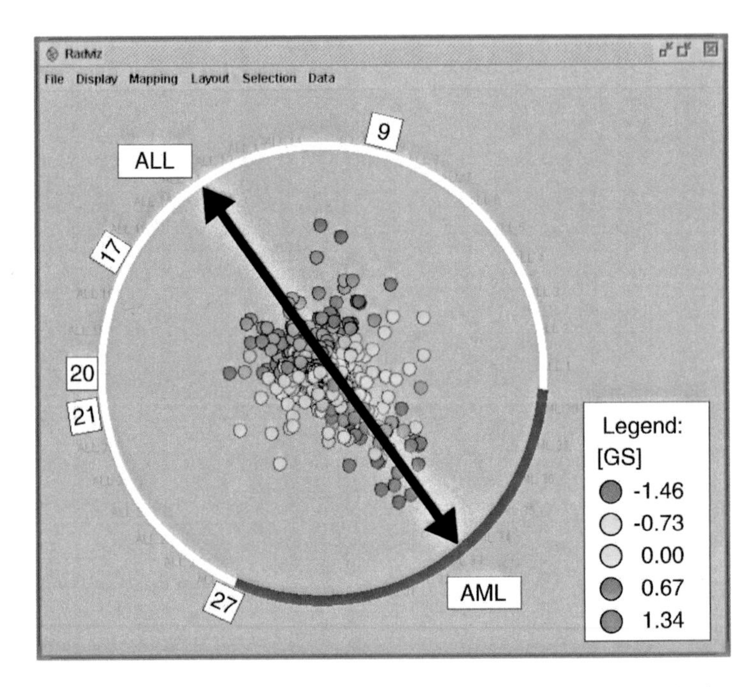

COLOR FIGURE 6.10 RadViz overview of the 6817 genes for the 38 patients in the training set.

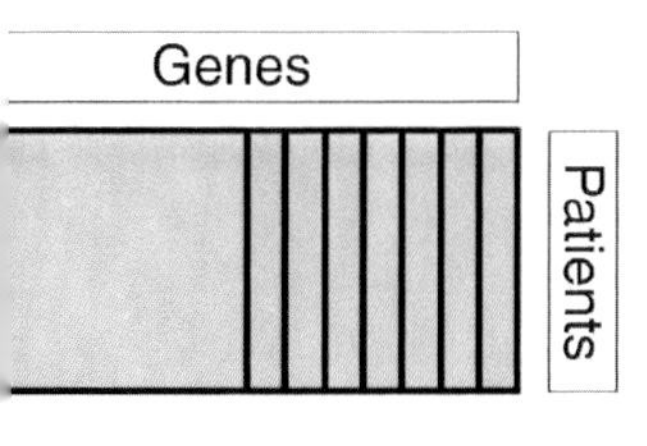

COLOR FIGURE 6.12 Pivoted data (in 6817 dimensions).

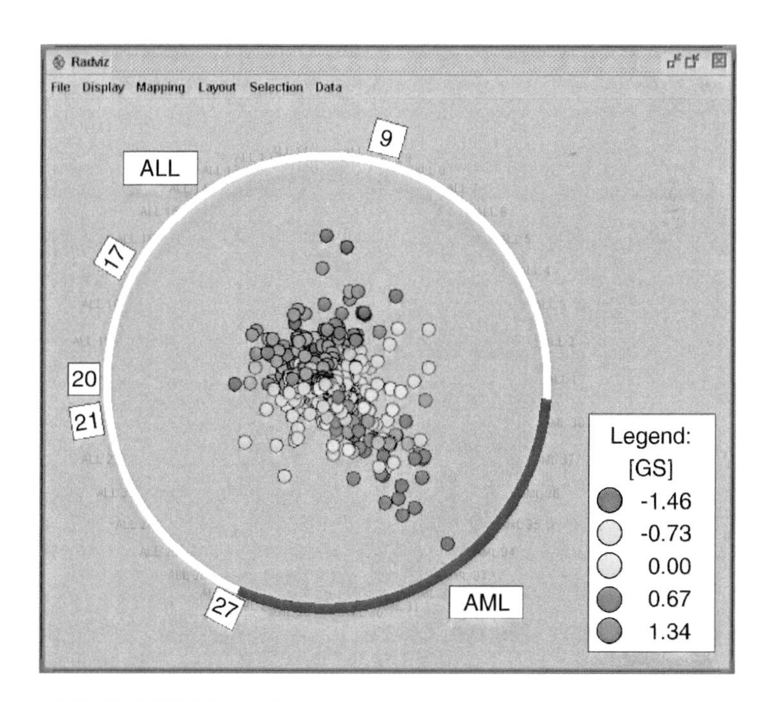

COLOR FIGURE 6.14 Selecting genes in lower-dimensional RadViz.

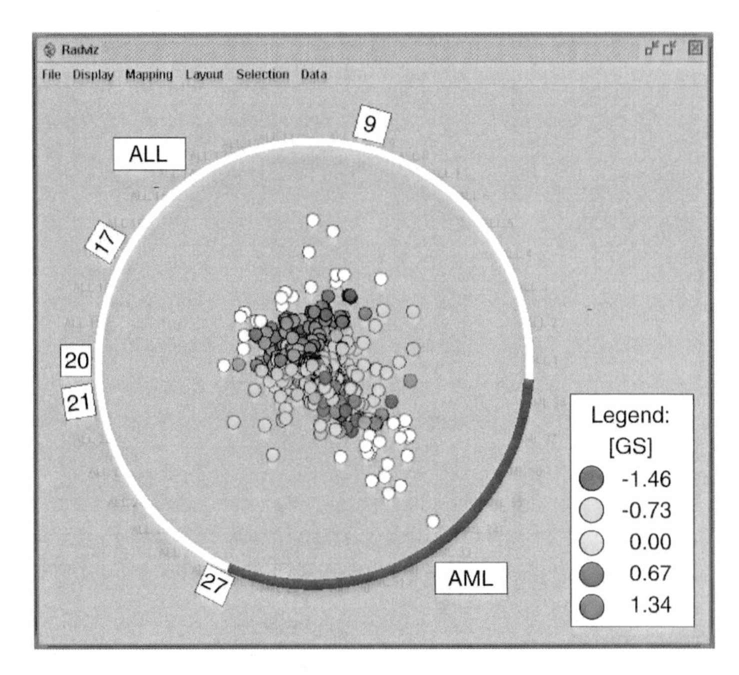

COLOR FIGURE 6.15 Selecting peripherally located genes.

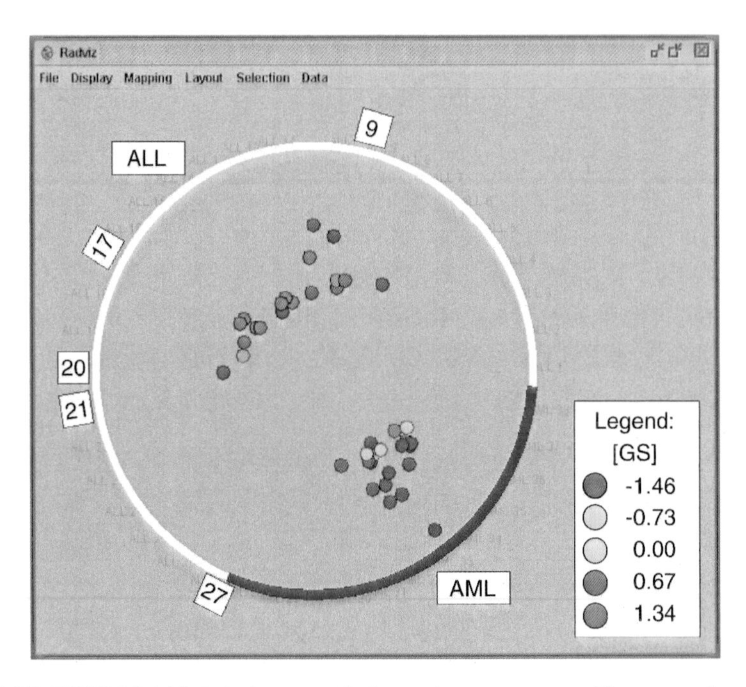

COLOR FIGURE 6.16 Deleting centrally located genes creates a 39 gene predictor set.

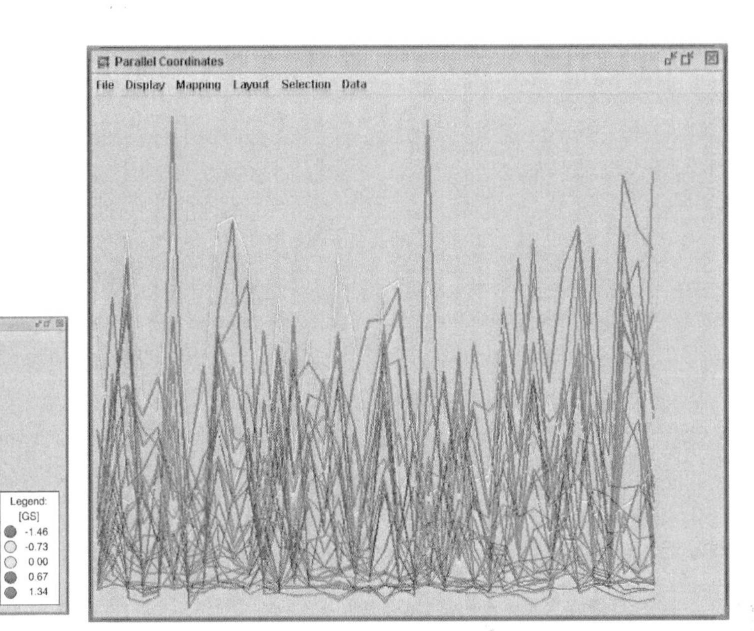

COLOR FIGURE 6.17 Parallel coordinate representation of a 39 gene predictor set.

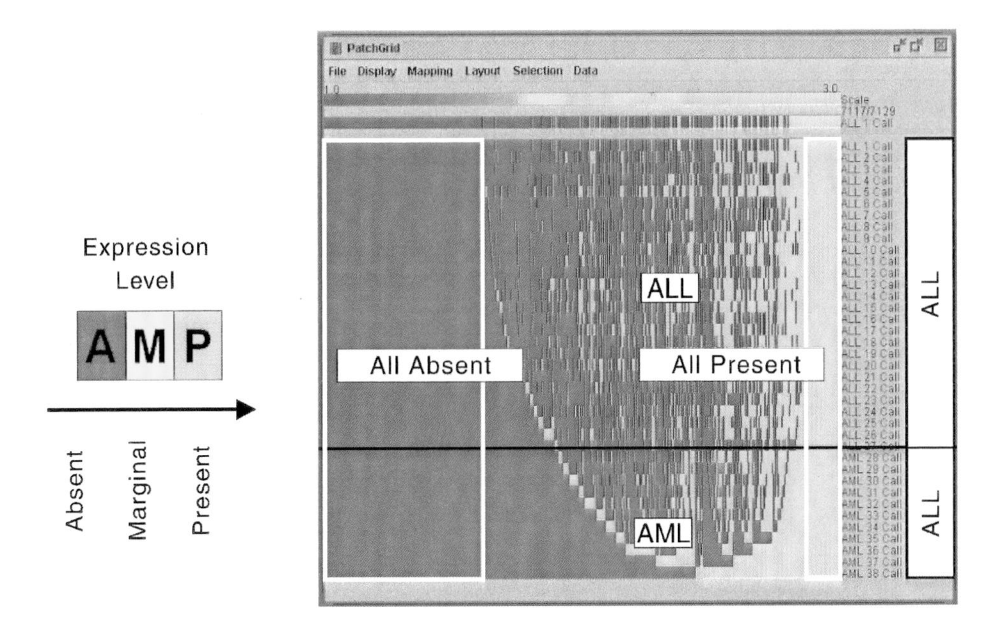

COLOR FIGURE 6.20 PatchGrid view of gene presence levels in all patients.

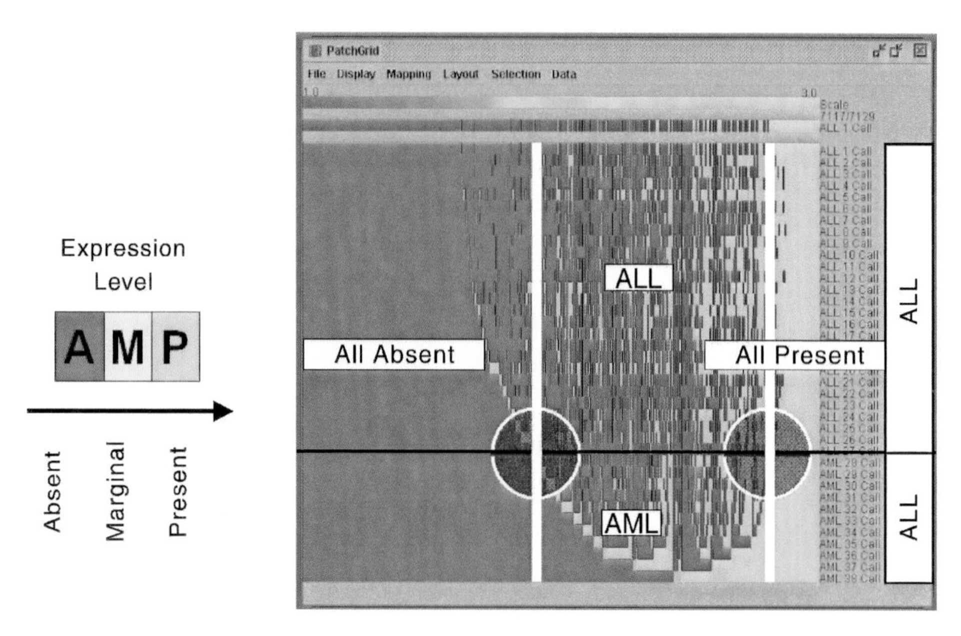

COLOR FIGURE 6.21 Boundary selections for better candidate gene selection.

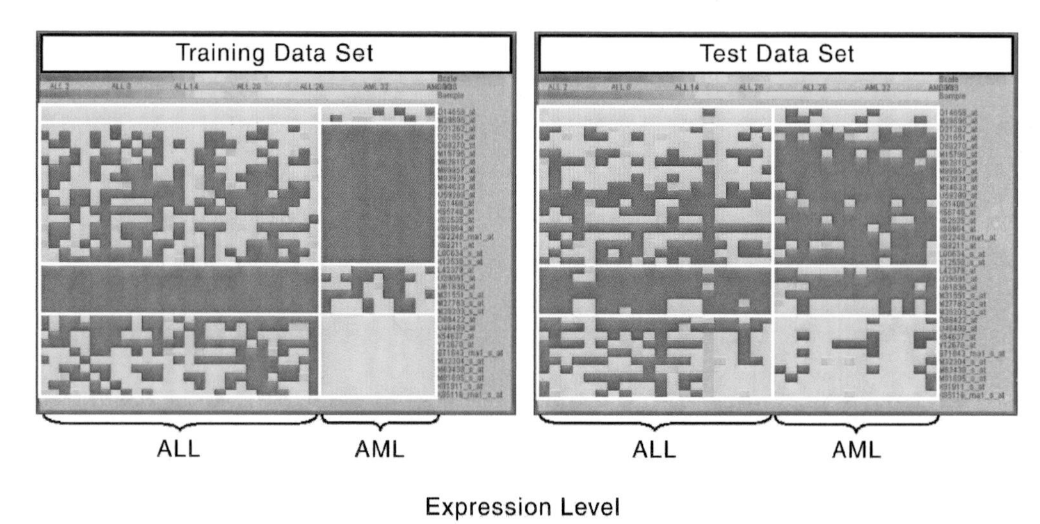

COLOR FIGURE 6.22 Predictor genes selected from the boundaries described in text (training data on left, test data on right).

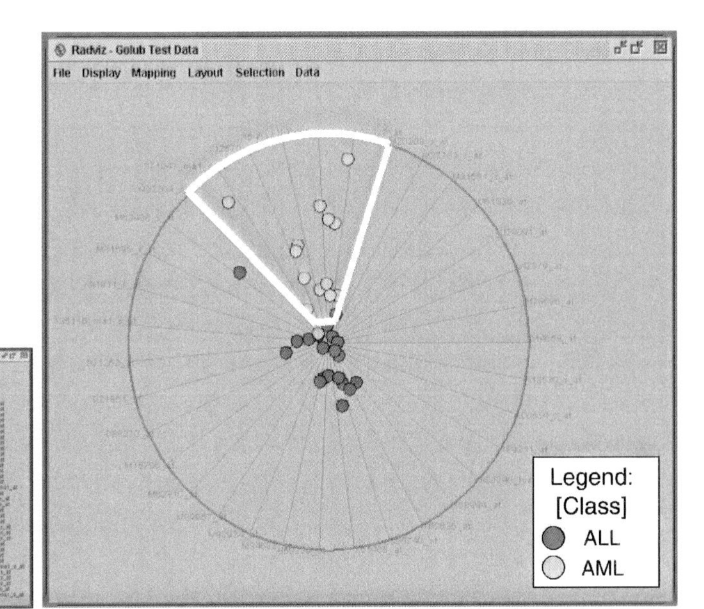

COLOR FIGURE 6.26 RadViz visual classifier applied to test dataset.

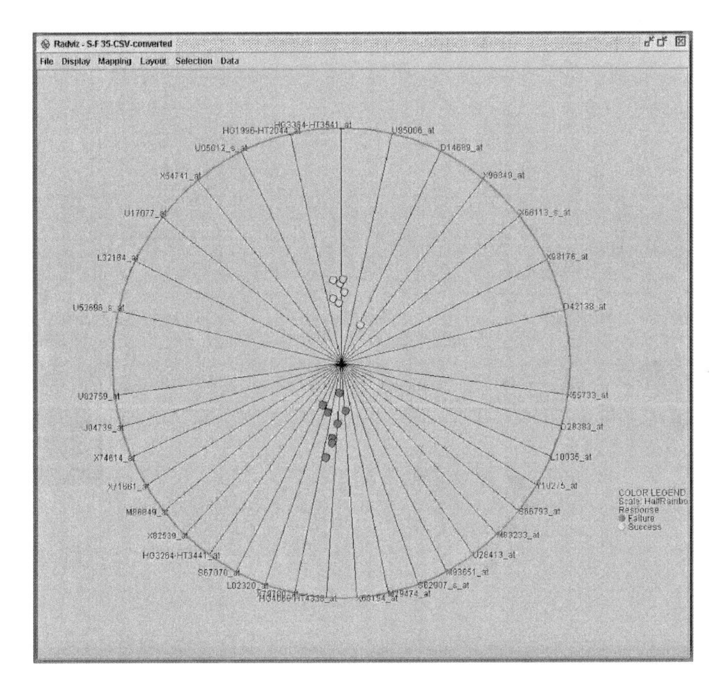

COLOR FIGURE 6.32 Classification by success or failure of treatment using 35 genes selected from the unfiltered set of 6817 genes by the class discrimination algorithm.

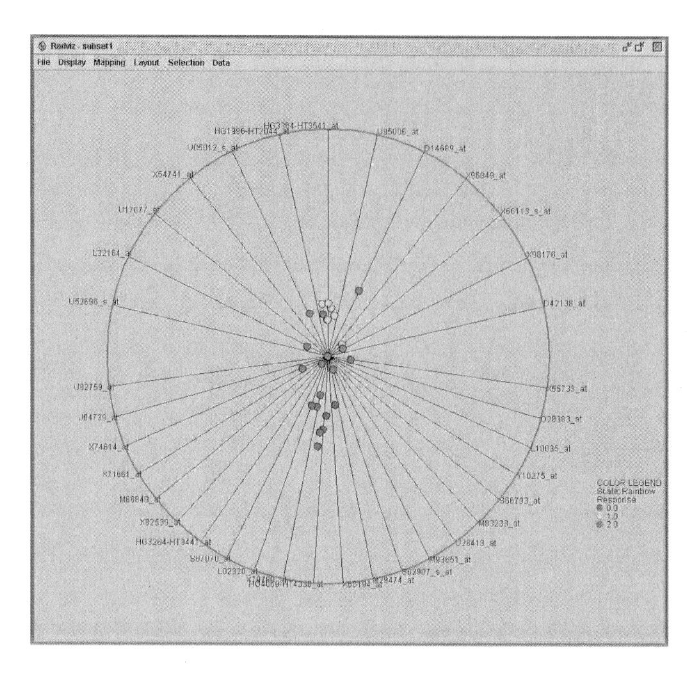

COLOR FIGURE 6.33 Prediction of unknown cases (0 is failure, 1 is success, and 2 is unknown).

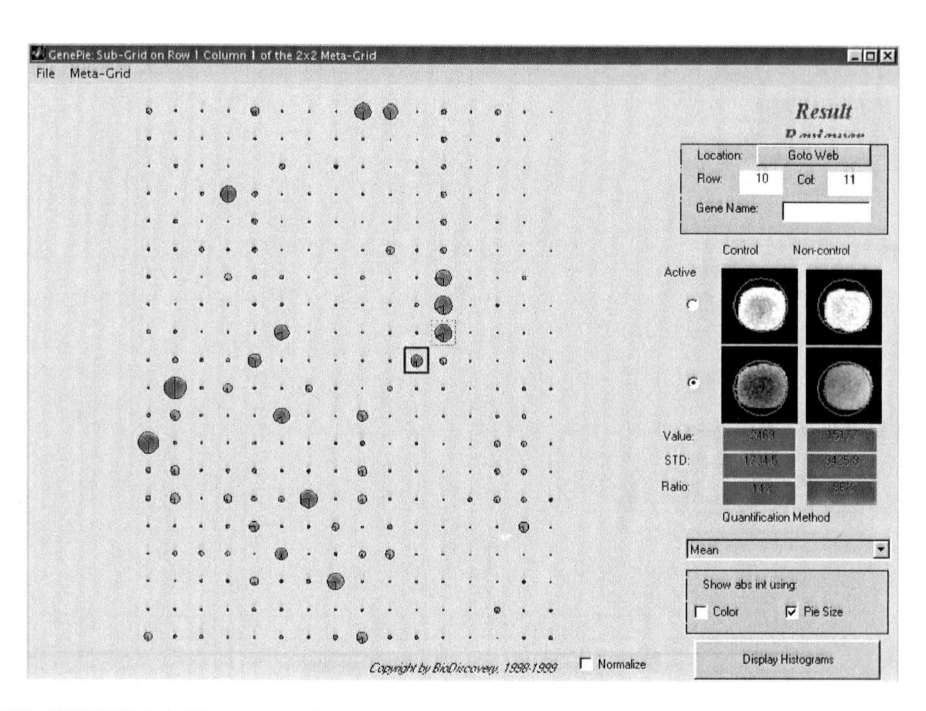

COLOR FIGURE 7.4 The time-series plot displays the expression levels of all genes across all experiments/files in the analysis. Experiments are plotted on the *x*-axis and expression levels of genes across all experiments are plotted on the *y*-axis. Clicking on a timeline, which finds the corresponding gene, or vice versa, monitors changes of expression levels over time.

We use visualizations not only as a means of portraying the results of analyses, but also as interactive tools for the exploration, manipulation, and analysis of the data and generation of subsequent results. Figure 6.4 outlines the fragment of a full exploration that we will limit ourselves to in this paper. The process depends on the data exploration task but does not vary significantly from the one described in the figure.

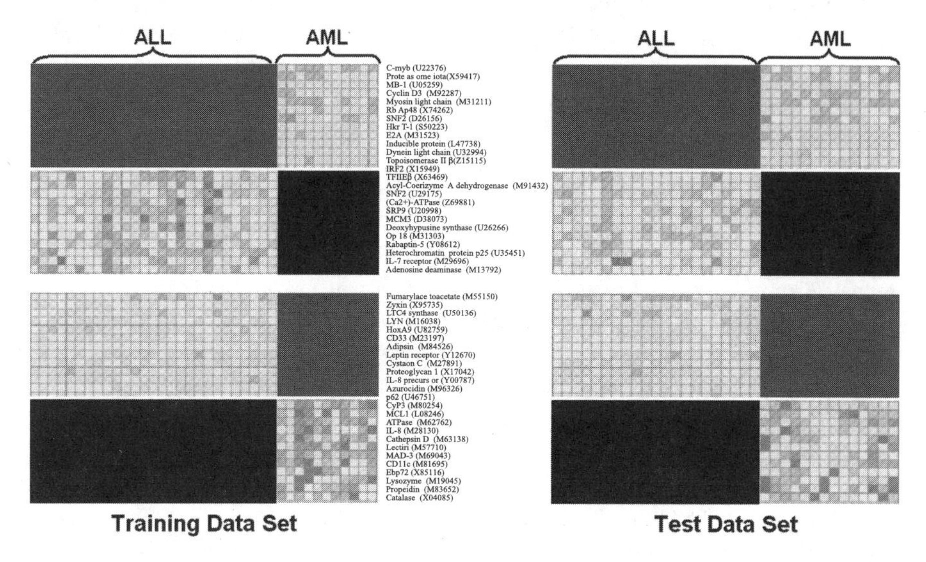

FIGURE 6.3 Gene expression grid. (From Golub, T.R. et al., *Science*, 286, 531–537, 1999. With permission.)

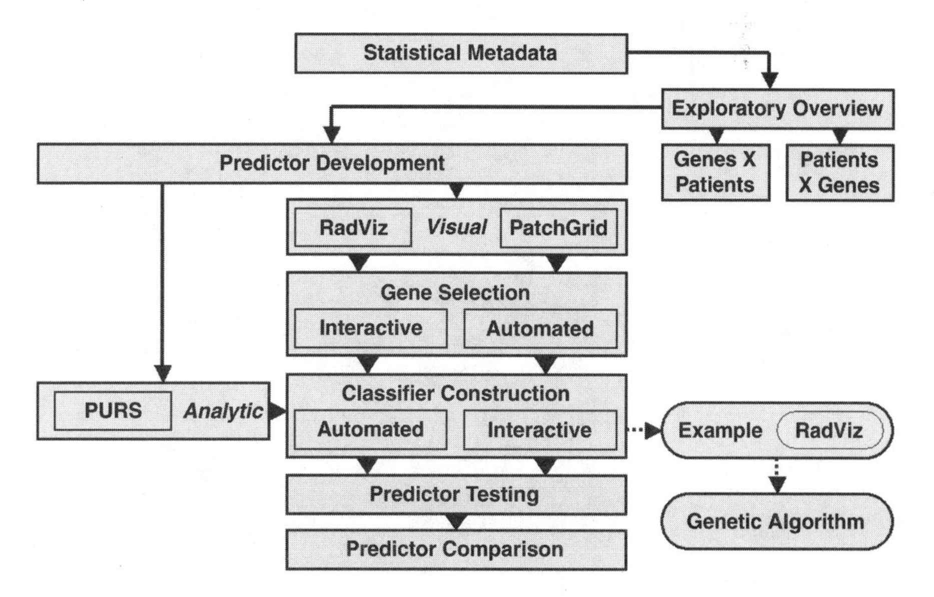

FIGURE 6.4 Fragment of the AnVil process applied to the Golub and Slonim dataset.

6.5.1 Metadata Exploration and Visualizations

In addition to a classic review of traditional statistical information about a dataset and the appropriate imputations of missing data, we create an overview of the statistics about the data in a PatchGrid™ (patent pending) metadata visualization (Figure 6.5). In this example, the statistics about gene expression levels for each of 38 samples in the training set are displayed in relative comparison. Standard statistical values, such as minimum, maximum, mean, mode, standard deviation, variance, skewness, kurtosis, and others, are given in this illustration. Values of each patient across each statistic are colored from relatively low to high on a black to white scale. Several patient samples stand out due to the presence of higher and/or lower values. Samples 9, 17, and 20 are pointed out in this display as samples that should be tracked for interesting or deviant contribution to results as analysis progresses.

If these samples significantly influence the results, the analysis might be repeated, omitting the samples. These records are tagged for tracking.

We then compute and display the Pearson correlation coefficient for gene expression level for all 6817 genes for each pair-wise set of patients for the 38×38 matrix of training set patients. In Figure 6.6, the correlations are colored on a range from Strong Negative as black and Strong Positive as white, with the expected 1:1 perfect correlation running down the diagonal. Other correlative measures can be used. Patient samples generally have a strong positive correlation, partly because all are from humans with leukemia and partly because "noisy" data composes over half of this relatively raw and unprocessed dataset. Samples from Patients 9, 20, and 21 stand out as having abnormally low correlations with other patients, and will be tracked.

This is typical of most discovery activities. Outliers pop up and need to be tagged for easy identification in downstream process activities. These may be meaningful.

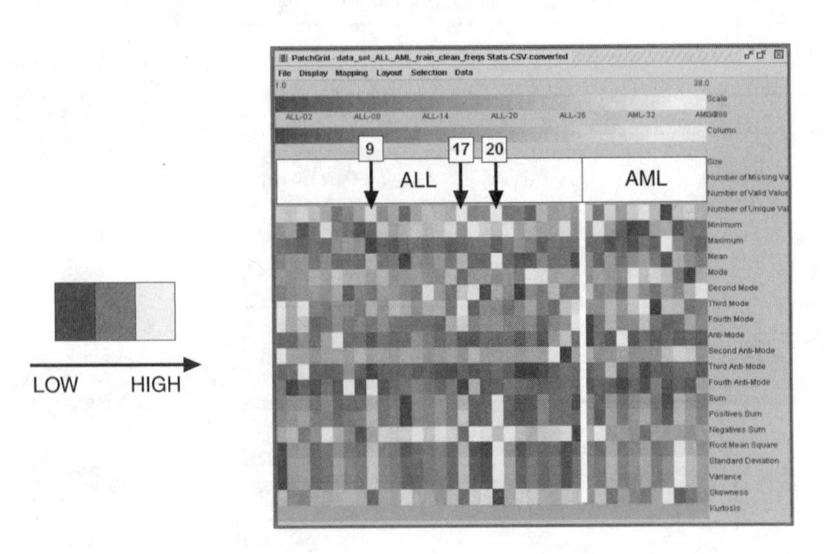

FIGURE 6.5 Metadata PatchGrid.

It is often these special cases that explain the anomalies produced in the discovery process.

Because this is Affymetrix data, we consider using the binned gene expression levels Absent, Marginal, and Present of the Affymetrix approach. These bins are created from rather complex algorithms empirically derived by Affymetrix and have not been well defined in the literature. Applying the Affymetrix binning alters the interpretation of the data, as potentially would any thresholding, filtering, or binning applied to the raw data. Figure 6.7 shows the results with patients 21 and 27 standing out in this visualization.

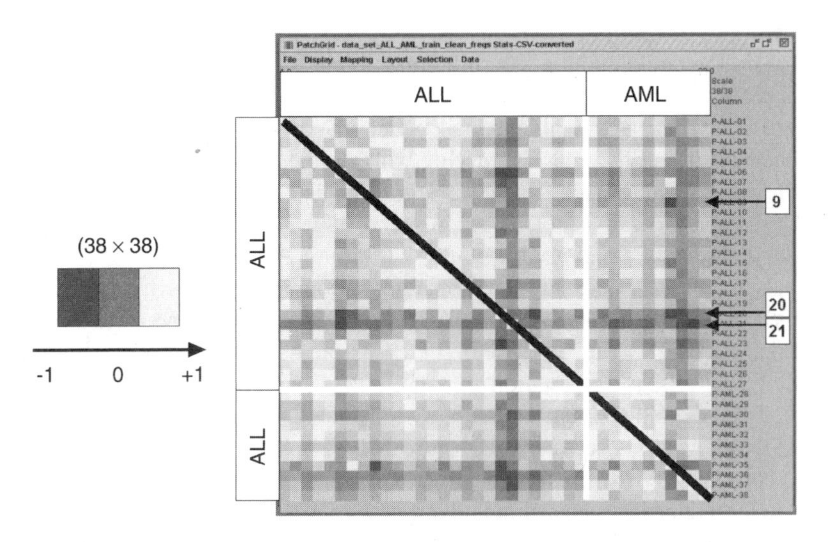

FIGURE 6.6 Pearson correlation PatchGrid.

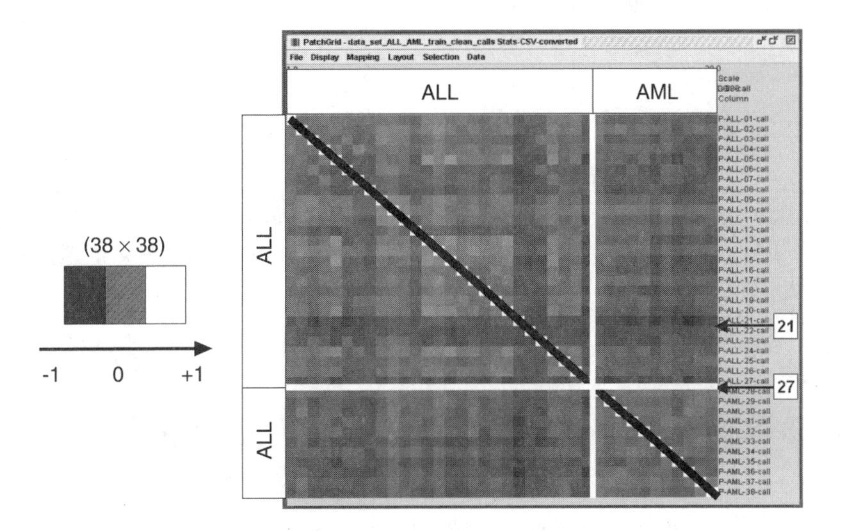

FIGURE 6.7 Pearson correlation PatchGrid on Affymetrix binning.

In summary, our metadata overview highlighted in Figure 6.8 leads us to track five samples to determine their impact on results and conclusions. Each visualization contributes to the pool of identified aberrant samples.

6.5.2 High-Dimensional Data Explorations

Our exploratory overview now proceeds to looking at all 6817 genes by the 38 patients in the training set using the RadViz™ (patent pending) display. RadViz displays data dimensions (columns; also known as parameters, variables, or fields) around a circle, and records (or rows) within the circle positioned according to a zero spring force resultant from virtual springs attached to the various dimensional anchors on the circle and the record points. Points with equal coordinates position at the center; points with one maximal value coordinate and all others minimal position at the dimension with maximal value; other points position within the circle with their position off center identifying which dimensions pull the most and thus have the largest values (see Reference 16 for details).

This display of all genes as records and 38 patients as columns, with ALL samples 1 through 27 and AML samples 28 through 38 simply ordered as given in the dataset about the axis, shows a global gene grouping having what appears to be an axis corresponding to the ALL—AML class axis. To determine whether this has meaning, we overlay the genes represented as points with a color scheme corresponding to a disease correlation measure calculated by Golub and Slonim (GS value). The GS value for each gene is the difference in mean, divided by the sum of standard deviations between ALL and AML samples (Figure 6.9). Negative GS values correlate to AML preferential expression and positive GS values correlate to ALL preferential expression of any given gene.

Looking at a distribution so colored in Figure 6.10, we see that the Magenta/Purple glyphs, with a positive GS value, do in fact visually segregate to the ALL axis and vice versa for the Red/Orange glyphs, with negative GS values, which are at

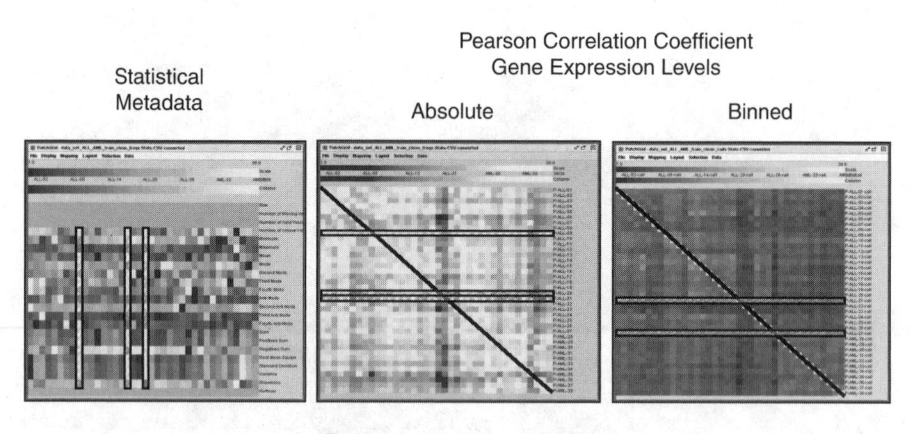

FIGURE 6.8 Summary of metadata analysis results. (See Color Figure 6.8 following page 82.)

the ALL pole of the axis. Hence, this demonstrates that a view of all available data can have meaning and, as later seen, can be used to pick predictor set genes.

6.5.3 Pivoted Data

Having looked at 6817 genes by 38 patients in Figure 6.10, we select all the genes (Figure 6.11), and pivot the table. This interchanges rows and columns and permits us to look at 38 patients by 6817 genes. This transposition, where records are now patients and columns are now genes, leads to a rather unorganized distribution of patients (Figure 6.12). Patient samples are positioned by RadViz in 6817-dimensional

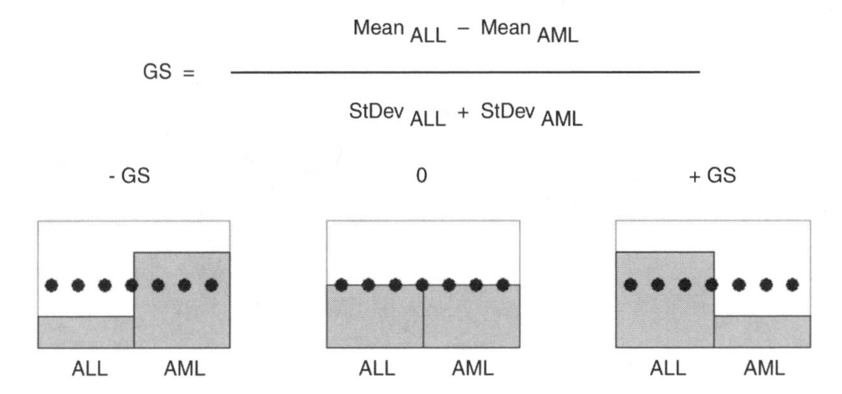

FIGURE 6.9 Golub and Slonim disease correlation measure.

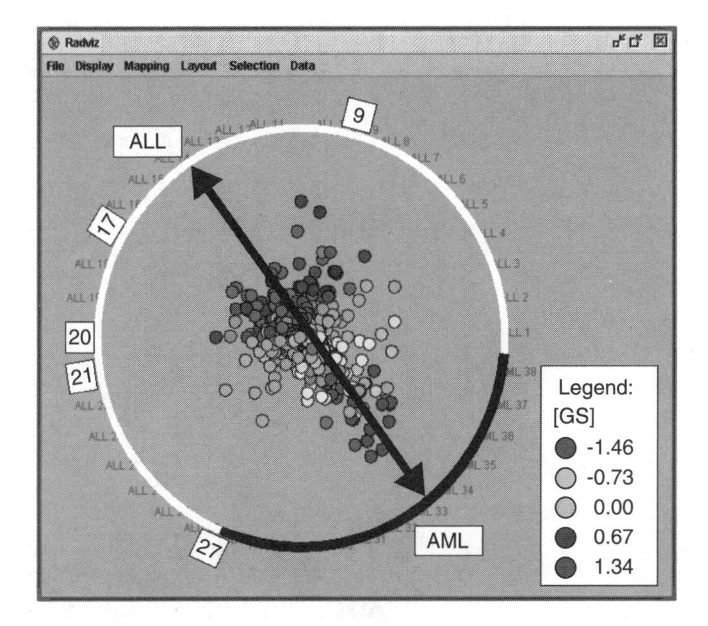

FIGURE 6.10 RadViz overview of the 6817 genes for the 38 patients in the training set. (See Color Figure 6.10 following page 82.)

space as defined by the total gene complement, which now comprise the columns in the data table. As genes are ordered as given in the output from the Affymetrix system, no meaningful data patterns would be evident unless the genes were ordered in some logical manner.

6.5.4 Clustering with RadViz

Meaning comes from this visualization (Figure 6.12) when a proprietary clustering algorithm is applied to order the genes in a rational manner (Figure 6.13). The training set shows very good discrimination between ALL and AML classes, with only one potential misclassification. As applied to the test set, there again appears

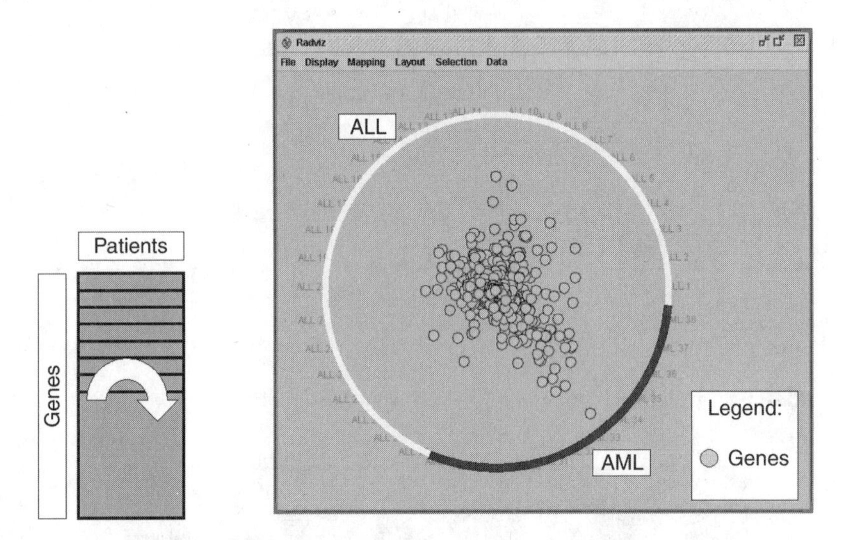

FIGURE 6.11 Pivoting genes and patients.

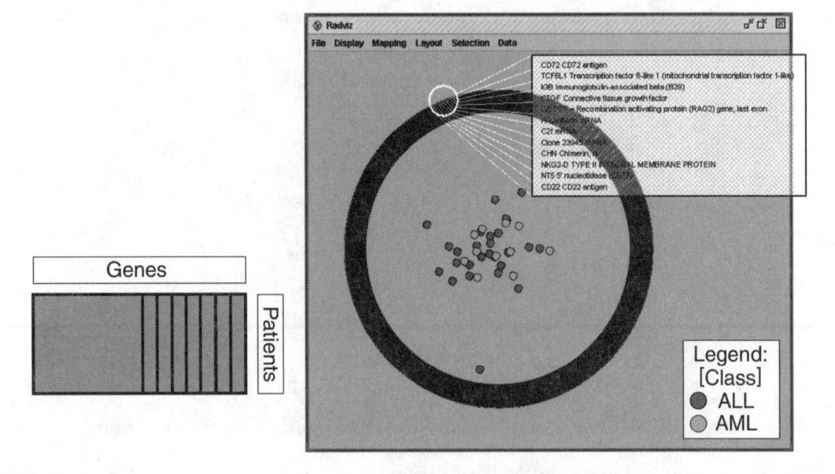

FIGURE 6.12 Pivoted data (in 6817 dimensions). (See Color Figure 6.12 following page 82.)

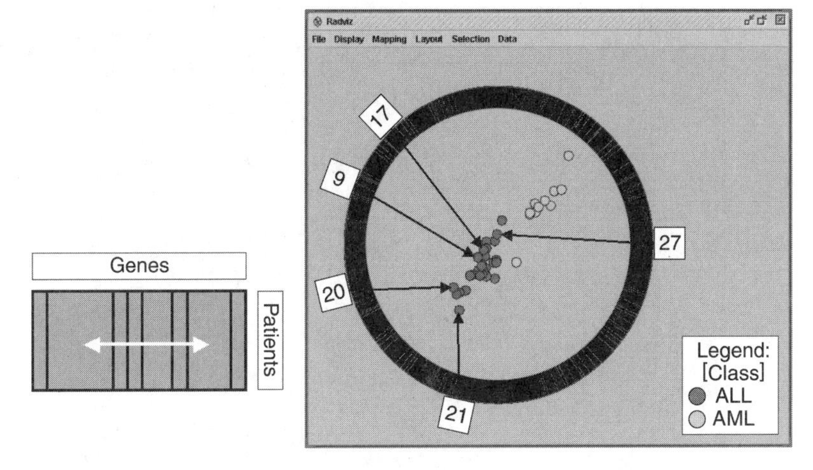

FIGURE 6.13 Clustered RadViz in 6817-dimensional space.

to be a polar discrimination between ALL and AML, although separation is not so distinct. The samples being tracked are shown, but have yet to come into play.

Working with the complete set of raw gene expression data, which has not been thresholded, filtered, or preprocessed in any way, it is surprising to see how much can be understood from working with the 6817 × 38 cell data matrix.

Viewing again (Figure 6.14) our earlier RadViz display of 38 patients by 6817 genes, recall that we confirmed that the genes at the periphery of the ALL—AML polar axis were indeed those genes whose expression ratio correlated with their respective location on the display.

6.5.5 Building a Predictor Set of Genes

By manually selecting (Figure 6.15) those peripherally located genes and deleting (Figure 6.16) the centrally located genes, we visually create a 39-gene predictor set for leukemia class distinction. These selected 39 genes are shown in Figure 6.17 using the familiar parallel coordinates display.[199]

As expected, those genes selected from the ALL zone in RadViz are highly expressed in ALL as seen in parallel coordinates, while those genes selected from the AML zone in RadViz are highly expressed in AML as seen in parallel coordinates (Figure 6.18).

6.5.6 Gene Expression Levels Explorations

To refine this selector set, we now look in more detail at the gene expression levels. In Figure 6.19, the 38 patients are arranged by rows and the 6817 genes by columns, with data presented as Affymetrix Absent/Marginal/Present call discretization of gene expression level. The columns are sorted via a proprietary algorithm (patent pending) that creates a dendrogrammatic view of genes absent in both diseases (Left), present in both diseases (Right), and differentially expressed in the intervening region.

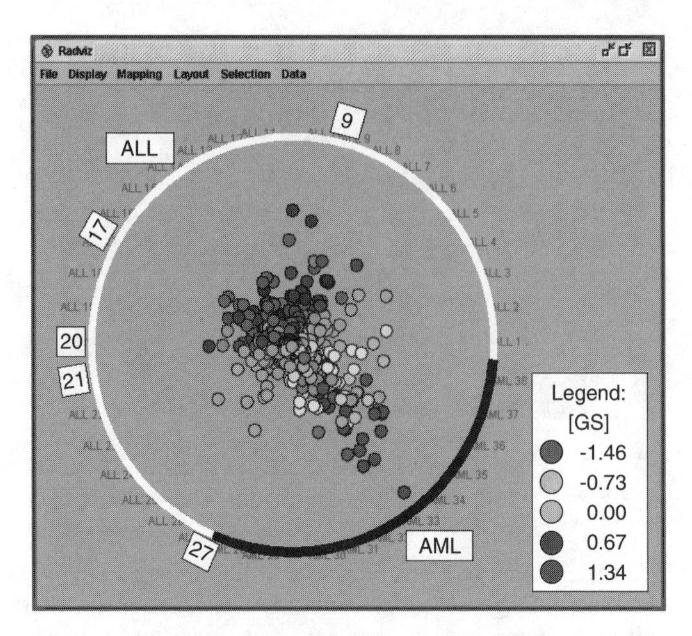

FIGURE 6.14 Selecting genes in lower-dimensional RadViz. (See Color Figure 6.14 following page 82.)

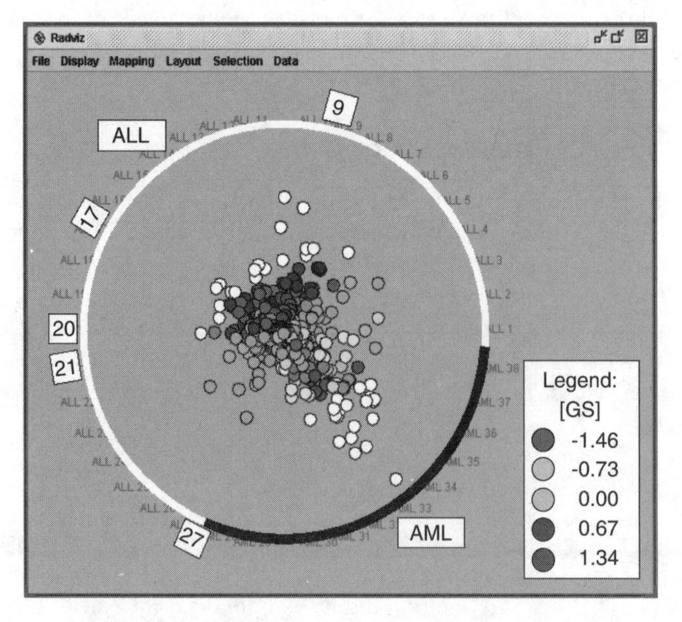

FIGURE 6.15 Selecting peripherally located genes. (See Color Figure 6.15 following page 82.)

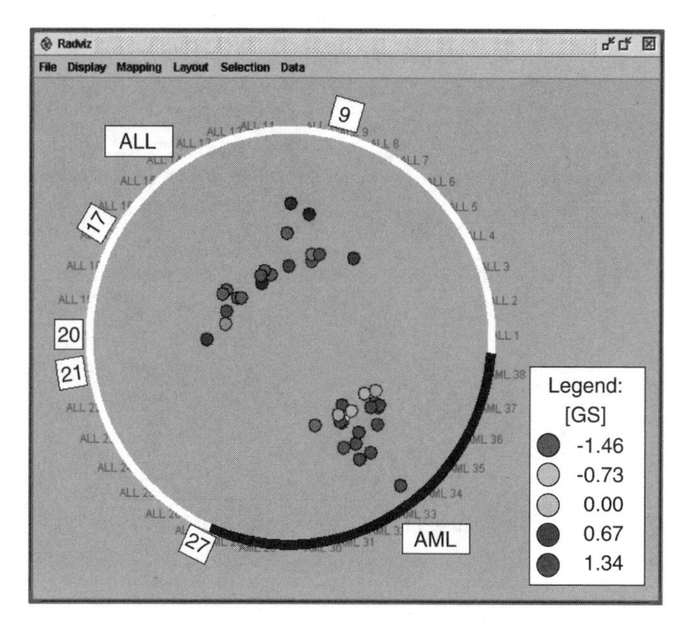

FIGURE 6.16 Deleting centrally located genes creates a 39-gene predictor set. (See Color Figure 6.16 following page 82.)

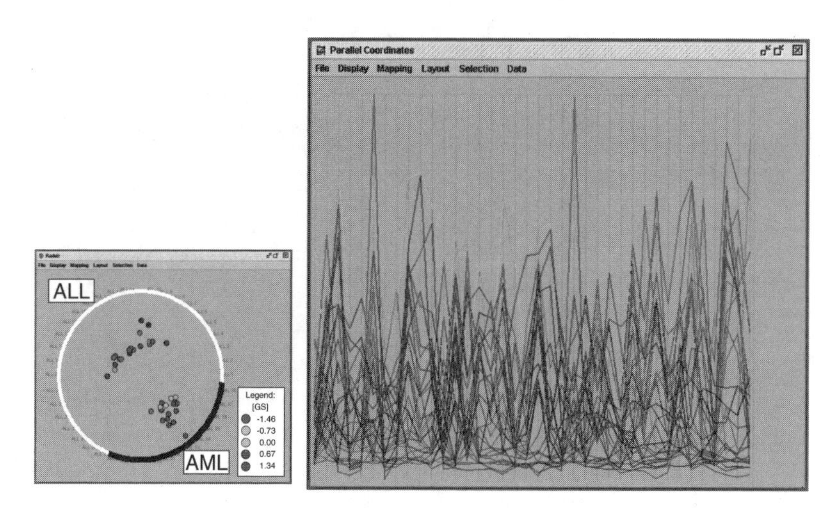

FIGURE 6.17 Parallel coordinate representation of a 39-gene predictor set. (See Color Figure 6.17 following page 82.)

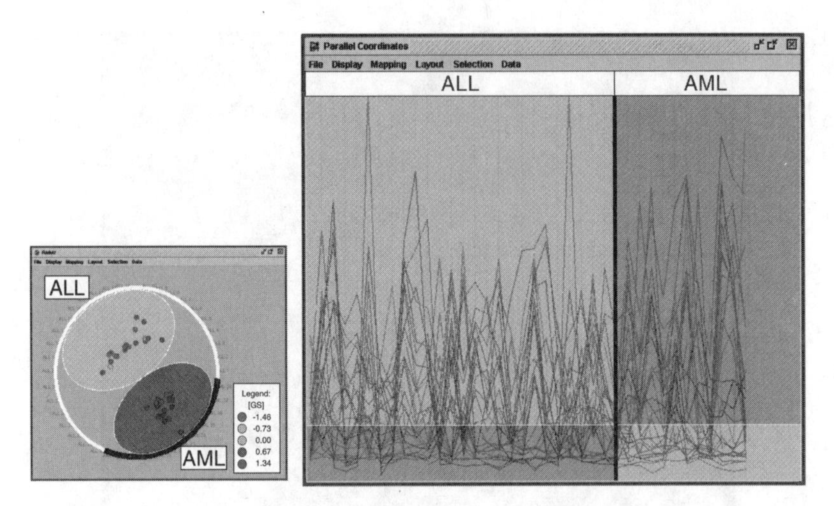

FIGURE 6.18 Parallel coordinate representation of expression levels of RadViz ALL/AML selections.

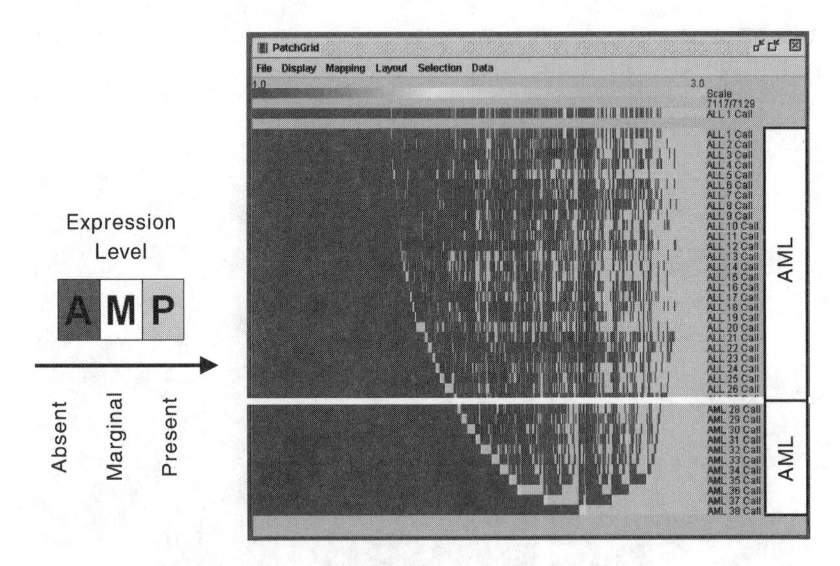

FIGURE 6.19 Gene expression levels for the 38 patients and all genes.

Genes always absent or always present in all patients in either leukemia class may be of interest in creating a predictor (Figure 6.20), but we are specifically interested in those that are positioned at the Absent/Present and ALL/AML boundaries. These are good candidates for predictor set selection (Figure 6.21).

Predictor set genes from the training data, selected from the Absent/Present, ALL/AML boundaries, are shown in the left panel of Figure 6.22. When applied to the test data, the right panel shows the visual output of the discrimination obtained. This view is analogous to that of Golub and Slonim shown in Figure 6.2.

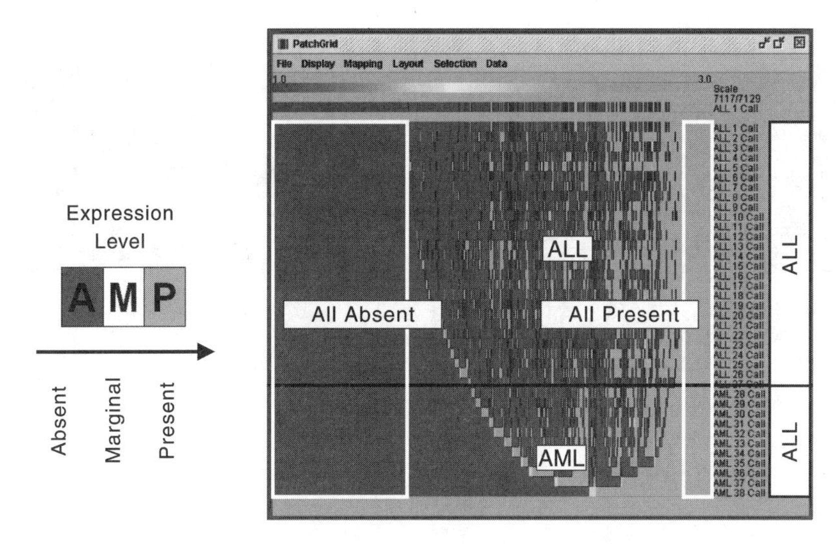

FIGURE 6.20 PatchGrid view of gene presence levels in all patients. (See Color Figure 6.20 following page 82.)

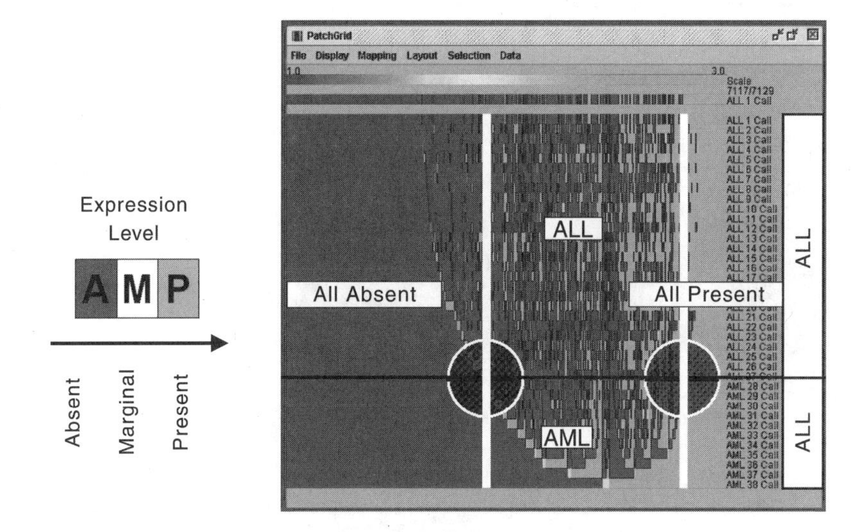

FIGURE 6.21 Boundary selections for better candidate gene selection. (See Color Figure 6.21 following page 82.)

As previously discussed, the ideal case would be a set of genes that is always "on" or "off" in a given disease state (Figure 6.23). The real world, however, allows us genes that are "on" in ALL and mostly "off" in AML, mostly "on" in ALL and "off" in AML, and the complementary case. These genes are those manually selected as a predictor set from training set data (Figure 6.24). As applied to the test set, the discrimination is less sharp, as was the case for Golub and Slonim's work, but again sufficient to classify the test set samples (Figure 6.22).

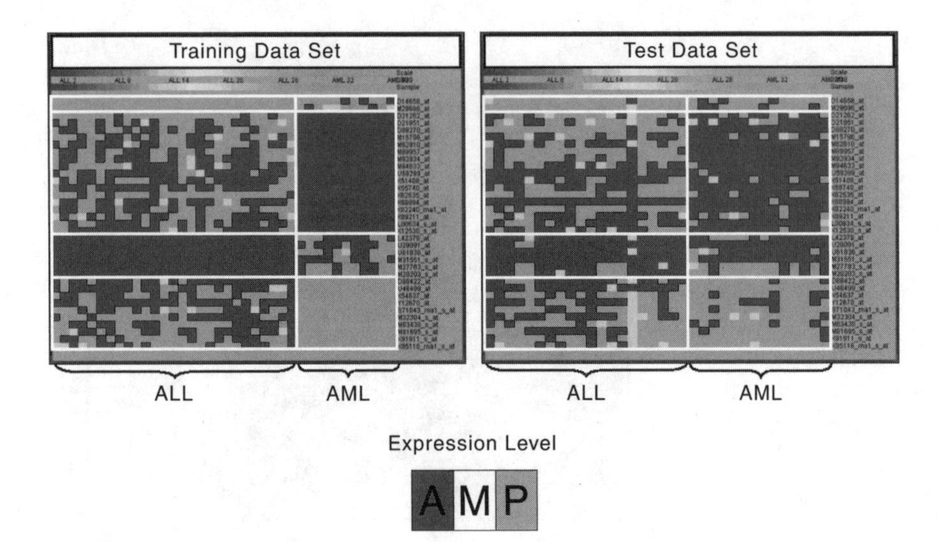

FIGURE 6.22 Predictor genes selected from the boundaries described in text (training data on left, test data on right). (See Color Figure 6.22 following page 82.)

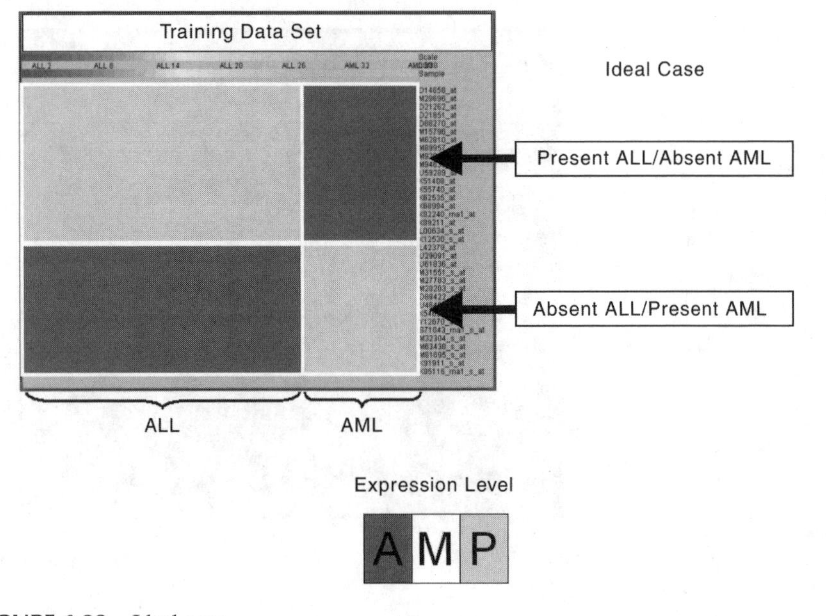

FIGURE 6.23 Ideal case.

6.5.7 A Visual Classifier

The 35 genes selected from the training set via PatchGrid (Figure 6.24) are now used to create another RadViz analytical visualization, again sorting the genes in a rational but unbiased manner (Figure 6.25). The white pie wedge denotes the visual classifier constructed from training set samples, with the samples being tracking depicted. Note that while the display utilizes lines and glyphs of sufficient dimensions

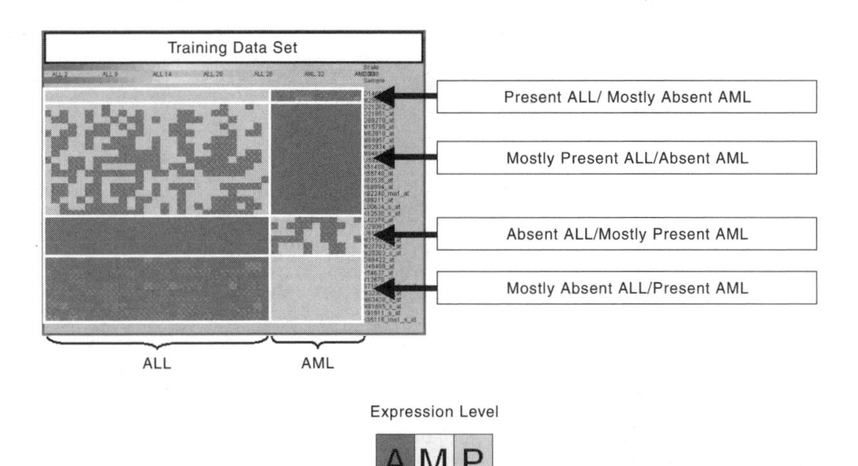

FIGURE 6.24 Genes manually selected as a predictor set from training set.

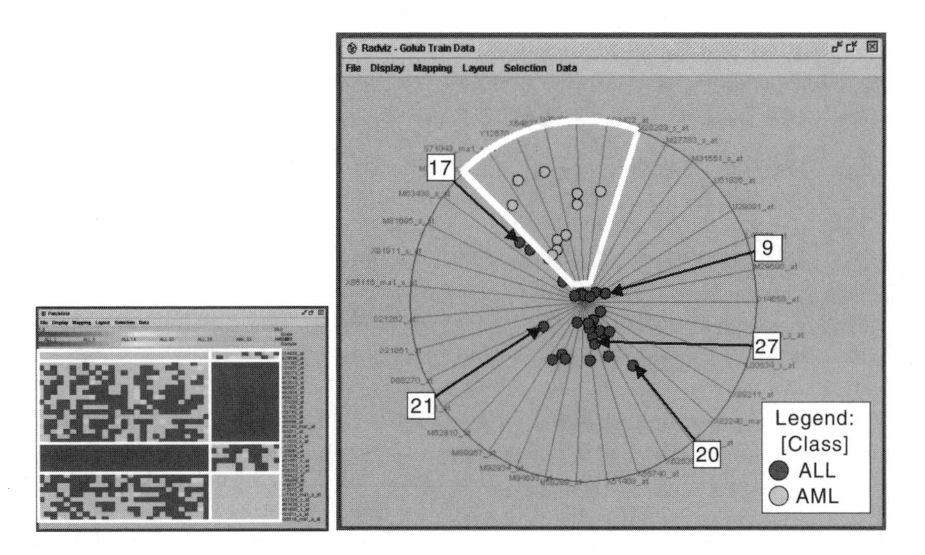

FIGURE 6.25 RadViz with ordered genes.

for visual evaluation, these features are mathematically minute. Hence, glyphs that appear to touch the border of a predictor boundary are mathematically *in* or *out* of the predictor set.

This same gene coordinate system and white pie wedge can now be used across the test dataset, correctly classifying all but one of the 34 test set samples (Figure 6.26). The performance of this predictor, as applied to the test set, exceeds that given by Golub and Slonim, as well as most others. A cleaner separation might come from application of additional means to further selectively reduce our gene set; an example of one such method is shown.

Having developed a classifier we now proceed to improve, refine, and test.

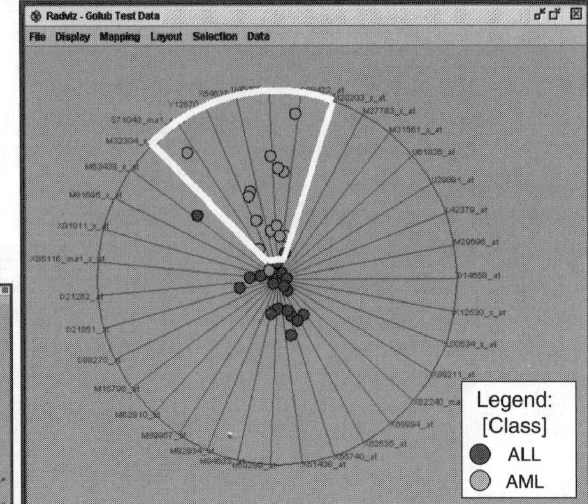

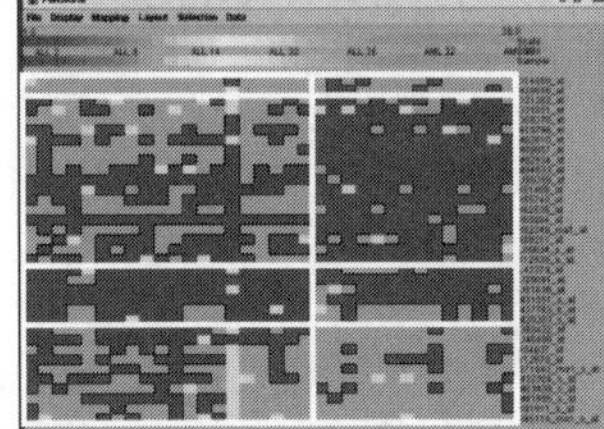

FIGURE 6.26 RadViz visual classifier applied to test dataset. (See Color Figure 6.26 following page 82.)

6.6 IMPROVING THE CLASSIFIER

6.6.1 Dimensional Reduction

An example of combined analytical and visual approaches to clustering uses RadViz to display the 38 training set patient samples in the three-dimensional space represented by three genes (Figure 6.27). These genes were obtained using a genetic algorithm to reduce a 76-gene predictor set constructed by a purely analytical method known as Principle Uncorrelated Record Selection (PURS). PURS reduces the dimensions of this dataset by eliminating genes that correlate with others, such that the final gene set is composed of genes uncorrelated to predetermined parameters. The three genes clearly distinguish ALL and AML clusters, as shown by the white border. ALL samples are, however, further subclustered in a meaningful way. These subclusters do, in fact, correspond to B-cell ALL and T-cell ALL cancer subtypes. Each of these genes needs to be explored for biological significance, especially in light of the possible random selection occurring (see Section 6.6.3).

The three genes shown are an EST, a B-cell associated membrane antigen, and leptin receptor, the latter gene the subject of speculative commentary by Golub and Slonim. It is very significant that the "driving force" separating B- and T-cell ALL samples is increased expression of a B-cell associated gene in the B-cell ALL samples. While two B-cell samples appear misclassified by this predictor, it is of interest to note that these samples are two of those that we have identified as suspect from our metadata overview. While such a predictor may have been rejected by other visual or analytical methods, our preliminary work to allow identification of outliers allows us to retain a predictor with strong biological implications.

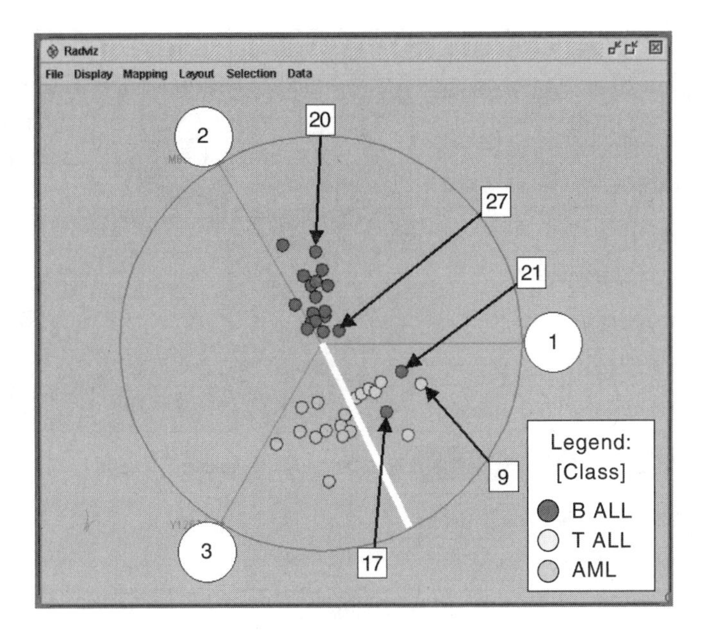

FIGURE 6.27 Three-gene RadViz patient display.

6.6.2 Classifier Evaluations

As a means of scoring or rating the efficacy of many different classifiers, we explored multiple analytical methods (NN, neural net; SVM, support vector machines; NB, naïve Bayes; LR, logistic regression; K, k-nearest neighbor with k = 3) (Figure 6.28). This chart depicts results for eight different classifiers, derived from both analytical and visual methods, as well as the Zyxin one-gene classifier identified by several different groups as the smallest sufficient classifier. Principal uncorrelated record selection (PURS) classifiers are found to perfectly classify both training and test samples for three of the five methods. We discuss this approach because it provides a control reference point for the numbers of genes in a valid classifier.

6.6.3 Random Gene Selection and PURS

This 76-gene analytic classifier, as depicted in the RadViz display in Figure 6.29, clearly demonstrates a delineation of ALL and AML samples in 76-dimensional space. This same boundary also correctly classifies the 34 samples from the test set (Figure 6.30). We now use this visualization as a predictor diagnostic of clinical treatment outcome.

Using the above visual classifier, we used PURS to produce a 76-gene set that predicts the success or failure of chemotherapy treatment outcome for acute myeloid leukemia (AML) (Figure 6.31). Known cases of successful treatment in black are distinguished from known cases where treatment failed in gray, with patients failing to go into remission.

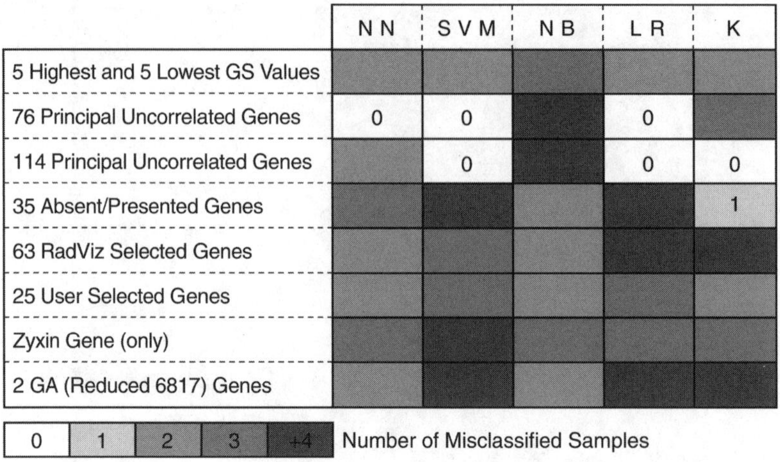

	NN	SVM	NB	LR	K
5 Highest and 5 Lowest GS Values					
76 Principal Uncorrelated Genes	0	0		0	
114 Principal Uncorrelated Genes		0		0	0
35 Absent/Presented Genes					1
63 RadViz Selected Genes					
25 User Selected Genes					
Zyxin Gene (only)					
2 GA (Reduced 6817) Genes					

0	1	2	3	+4	Number of Misclassified Samples

FIGURE 6.28 Classifier comparisons.

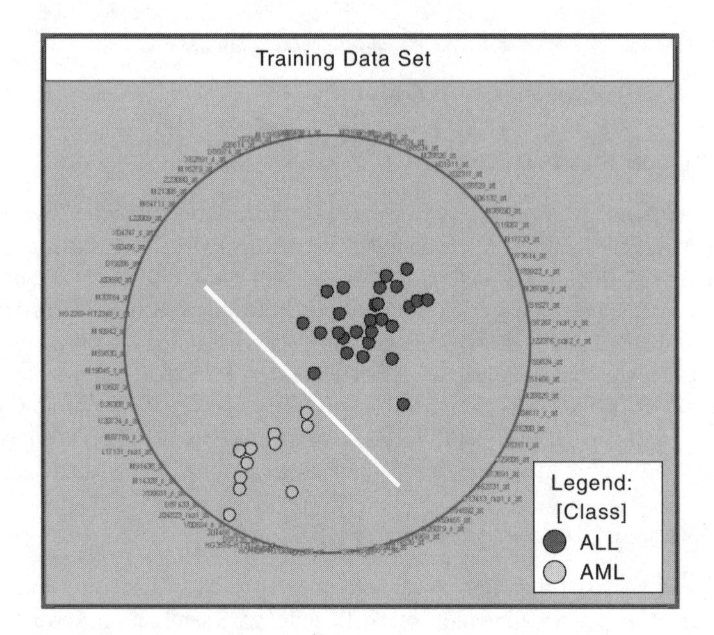

FIGURE 6.29 The 76-gene PURS classifier on the training dataset.

Based on statistical computations, 72 patients can be perfectly classified by 72 genes, provided the gene expression values are sufficiently uncorrelated (*read* random enough). This explains why the 76-PURS gene set works so well in classification. In fact, there will be many ways to do this. Given any partition of the subjects into classes, this selection of variables can classify them successfully.

In that case, what is the number of biologically significant genes within that subset? This depends on the number of highly correlated genes within that 76-gene collection. For example, in the Golub and Slonim dataset, there are over 100 gene

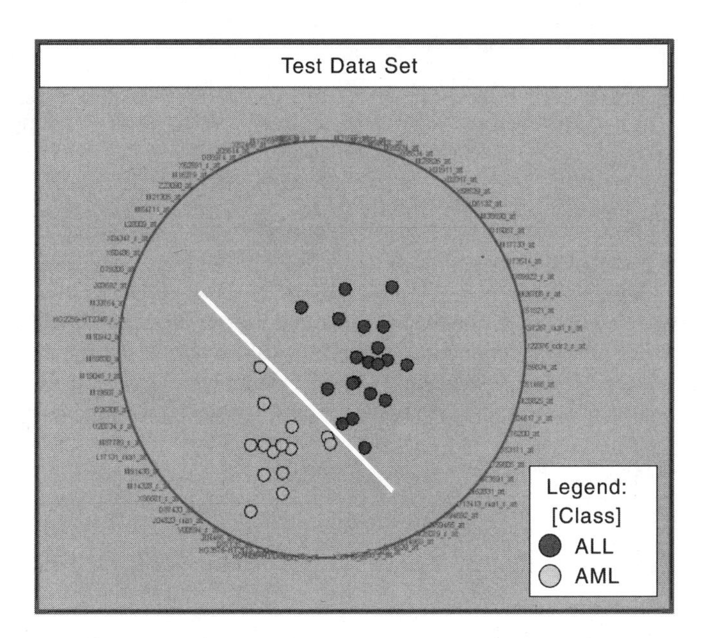

FIGURE 6.30 The 76-gene PURS classifier on the test dataset.

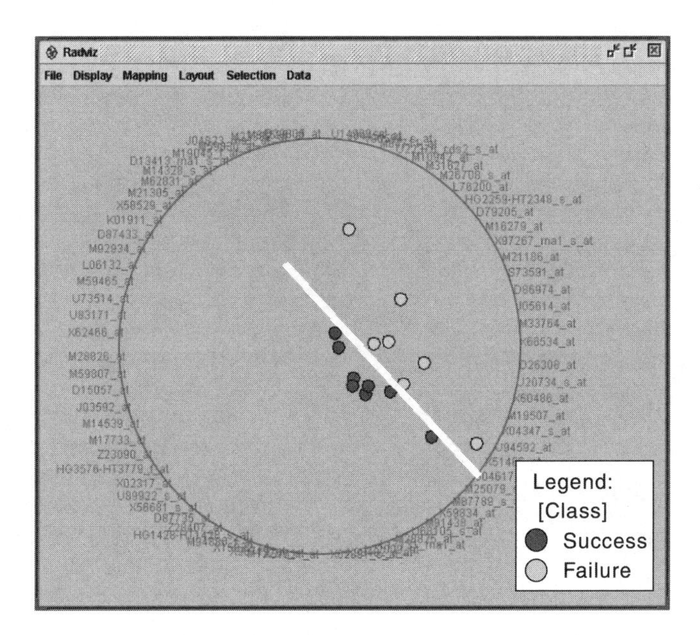

FIGURE 6.31 The 76-gene set predictor for the success or failure of chemotherapy treatment outcome for acute myeloid leukemia (AML).

pairs with correlation higher than 0.95 across the 38 training samples. Hence, the genes in any patient sample are highly correlated. The likelihood of the PURS gene set not being reducible is thus very high. Therefore, it makes better sense to explore another collection of genes.

6.6.4 Improving the Classifier

Using a proprietary layout algorithm involving success/failure class discrimination on the original dataset, we reduce the gene collection from 6817 to 35. Figure 6.32 displays this new gene collection. Note the strong separation.

We now use this classifier as a predictor. A further set of ten AML patients, for whom treatment outcome is not available, is plotted on this display (Figure 6.33, circles colored in magenta) to predict treatment outcome. However, patient data cannot be obtained to confirm our prediction. We predict that for four patients, chemotherapy would be successful, for four not successful, and for one unknown. It is clear in this case that statistical validation is necessary, as is the next step.

This selection of 35 genes resulted in the perfect classification of the success or failure of treatment using hold-one-out validation with all the following classifiers:

- Naïve Bayes
- Support vector machine
- One nearest neighbor
- Three nearest neighbors
- Logistic regression
- Neural network

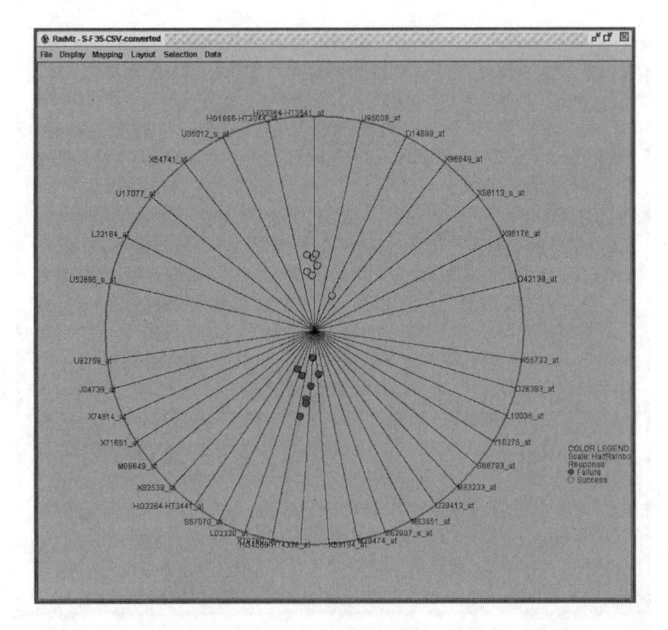

FIGURE 6.32 Classification by success or failure of treatment using 35 genes selected from the unfiltered set of 6817 genes by the class discrimination algorithm. (See Color Figure 6.32 following page 82.)

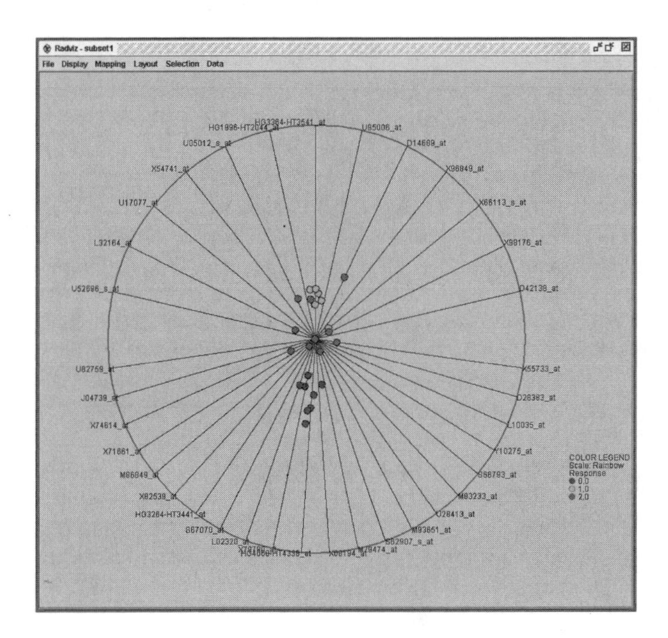

FIGURE 6.33 Prediction of unknown cases (0 is failure, 1 is success, and 2 is unknown). (See Color Figure 6.33 following page 82.)

It is interesting to note that of the dozens of research groups who have published analyses of the Golub and Slonim dataset, including the original authors, none have been able to present a gene predictor set diagnostic of clinical treatment outcome.

While a survival prognosis predictor has been generated for lymphoma,[1] a gene predictor set for prediction of the success or failure of chemotherapy treatment is clearly a unique contribution.

6.7 COMPARISON OF RESULTS WITH RESULTS FROM CAMDA

The December 2000 *Critical Assessment of Techniques for Microarray Data Analysis* (*CAMDA 2000*) meeting presented the Golub and Slonim dataset as a common starting point for comparison of microarray analytical methods. Of the 32 presentations evaluating the dataset, few presented the genes composing their derived class predictor sets.

We have used a visualization to depict, in classical parallel coordinates style, the relationship between gene predictor sets among the published groups and AnVil (Figure 6.34). AnVil's set is the union of six separate sets derived from different methods. As groups variably applied preprocessing (including threshold, filter, and other techniques) and different gene selection algorithms in construction of predictor sets, common as well as unique genes were identified in each predictor set.

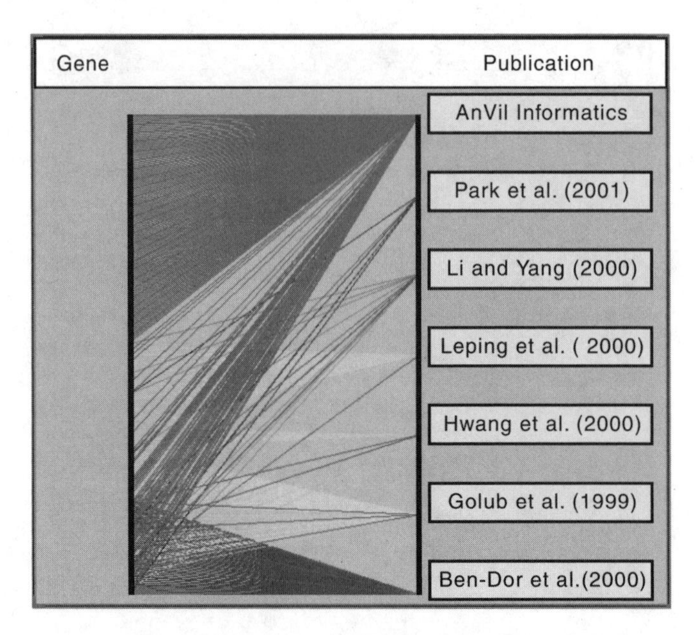

FIGURE 6.34 Parallel coordinate representation of the relationship between gene predictor sets among the published CAMDA 2000 groups and AnVil.

Using further analytical and visualization techniques to generate and analyze predictor sets, we can take this even further. We can categorize and rank the various methods and advise on appropriate gene selections for drug targets and diagnostic applications. As mentioned, every sufficient method employed by a capable analyst will yield some results. Being able to judge the quality and significance of results is paramount in moving forward to utilize these results in discovery and development programs.

In Figure 6.35 we overlay the top three genes identified in common in the seven gene sets presented, with four of seven sets containing all three genes, two of seven containing only two of the genes and one set containing one of the three top genes. This example illustrates the point that the lowest hanging fruit will be readily identified by many methods, and that judging the accuracy and reliability of results will require the self-critical evaluation that we employ.

6.8 CONCLUSION

We can basically dispel the myth that any data mining algorithm will work on any dataset. Believing that the latest XYZ algorithm will give the best and truest answers really only provides "an" answer. That answer needs to be interpreted. It is somewhat like taking one's temperature. The results are but one measure possibly indicative of much more. Taken to its extreme, sick data mining looks like healthy

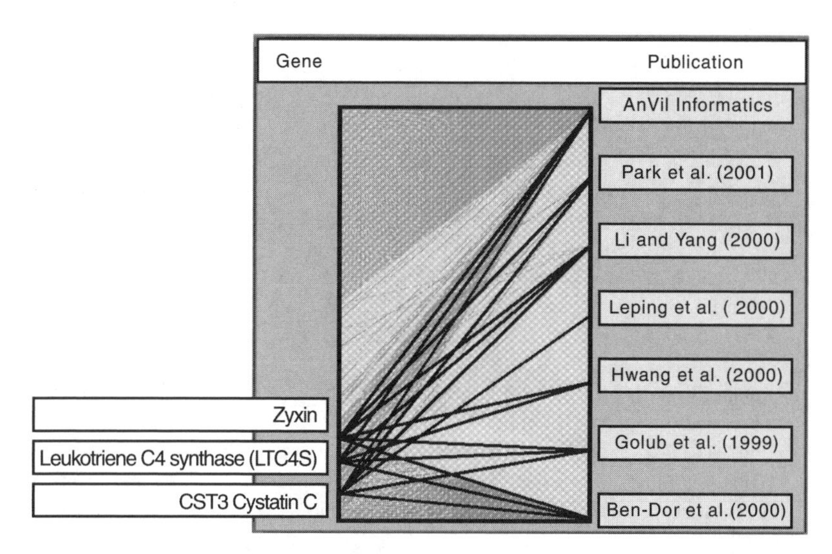

FIGURE 6.35 Top three common genes.

data mining if one does not seek the advice of a professional. Shrink-wrapped software is not a solution.

This chapter has demonstrated how high-dimensional visualization of massive microarray datasets can reveal valuable clustering relationships, even before filtering, thresholding, and other preprocessing, using the data of Golub and Slonim.

We identified several suspect patient samples that were subsequently shown to be falsely misclassified samples in a three-gene predictor set that otherwise classified B- and T-cell ALL and AML based on the influence of a B-cell associated gene. This biologically important result may have been discarded by others using methods not evaluating data quality along with classifier prediction. The analysis of limited chemotherapy treatment outcome data also yielded a 76-gene predictor for remission success and failure — a finding unique among the dozens of presented analyses of the Golub and Slonim dataset.

Traditional analysis and visualization methods are unable to look at the entire dataset as a whole due to limitations of existing software tools. Assumptions made in the reduction of data dimensions may also remove or obscure data relationships and candidate genes, thereby resulting in weaker classifiers and predictors and loss of valuable targets. Our approach preserves the value and meaning inherent in the full dataset by creating global data and meta-statistical overviews that reveal major data patterns and identify aberrant samples that may bias results.

Early identification of clustering potentials and outlier samples avoids mistakes that are costly in terms of time and expense and thus slow or misdirect the drug discovery process. High-dimensional data visualization increases the likely identification of such clusters and outliers.

REFERENCES

1. Alizadeh, A.A., Eisen, M.B., Davis, R.E., Ma, C., Lossos, I.S., Rosenwald, A., Boldrick, J.C., Sabet, H., Tran, T., Yu, X., Powell, J.I., Yang, L., Marti, G.E., Moore, T., Hudson, J., Jr., Lu, L., Lewis, D.B., Tibshirani, R., Herlock, G., Chan, W.C., Greiner, T. C., Weisenburger, D.D., Armitage, J.O., Warnke, R., Levy, R., Wilson, W., Grever, M.R., Byrd, J.C., Botstein, D., Brown, P.O., and Staudt, L.M., Distinct types of diffuse large B-cell lymphoma identified by gene expression profiling, *Nature*, 403, 503–511, 2000.
2. Alter, O., Brown, P.O., and Botstein, D., Singular value decomposition for genome-wide expression data processing and modeling, *Proc. Natl. Acad. Sci. U.S.A.*, 97, 10101–10106, 2000.
3. Bassett, D.E., Jr., Eisen, M.B., and Boguski, M.S., Gene expression informatics: it's all in your mine, *Nat. Genet.*, 21, 51–55, 1999.
4. Cho, R.J., Campbell, M.J., Winzeler, E.A., Steinmetz, L., Conway, A., Wodicka, L., Wolfsberg, T.G., Gabrielian, A.E., Landsman, D., Lockhart, D.J., and Davis, R.W., A genome-wide transcriptional analysis of the mitotic cell cycle, *Mol. Cell.*, 2, 65–73, 1998.
5. Chu, S., DeRisi, J., Eisen, M., Mulholland, J., Botstein, D., Brown, P.O., and Herskowitz, I., The transcriptional program of sporulation in budding yeast, *Science*, 282(5389), 699–705, 1998.
6. Cvek, U., Gee, A., Hoffman, P., Pinkney, D., Trutschl, M., Zhang, H., Marx, K., and Grinstein, G., Data Mining of Yeast Functional Genomics Data Using Multidimensional Analytic and Visualization Techniques, Drug Discovery Technology 1999, Boston, MA, August 1999.
7. De Risi, J.L., Iyer V.R., and Brown, P.O., Exploring the metabolic and genetic control of gene expression on a genomic scale, *Science*, 278, 680–686, 1997.
8. Diehn, M., Eisen, M.B., Botstein, D., and Brown, P.O., Large-scale identification of secreted and membrane-associated gene products using DNA microarrays, *Nat. Genet.*, 25, 58–62, 2000.
9. Eisen, M.B., Spellman, P.T., Brown, P.O., and Botstein, D., Cluster analysis and display of genome-wide expression patterns, *Proc. Natl. Acad. Sci. U.S.A.*, 95(25), 14863-14868, 1998.
10. Ferea, T.L., Botstein, D., Brown, P.O., and Rosenzweig, R.F., Systematic changes in gene expression patterns following adaptive evolution in yeast, *Proc. Natl. Acad. Sci. U.S.A.*, 96(17), 9721–9726, 1999.
11. Gasch, A.P., Spellman, P.T., Kao, C.M., Carmel-Harel, O., Eisen, M.B., Storz, G., Botstein, D., and Brown, P.O., Genomic expression programs in the response of yeast cells to environmental changes, *Molec. Biol. Cell*, 11(12), 4241–4257, 2000.
12. Girman, C.J., Cluster Analysis and Classification Tree Methodology as an Aid to Improve Understanding of Benign Prostatic Hyperplasia, Ph.D. thesis, Department of Biostatistics, University of North Carolina, Chapel Hill, 1994.
13. Golub, T.R., Slonim, D.K., Tamayo, P., Huard, C., Gaasenbeek, M., Mesirov, J.P., Coller, H., Loh, M.L., Downing, J.R., Caligiuri, M.A., Bloomfield, C.D., and Lander, E.S., Molecular classification of cancer: class discovery and class prediction by gene expression monitoring, *Science*, 286, 531–537, 1999.
14. Gray, N.S., Wodicka, L., Thunnissen, A.W.H., Norman, T.C., Kwon, S., Espinoza, F.H., Morgan, D.O., Barnes, G., LeClerc, S., Meijer, L., Kim, S., Lockhart, D.J., and Schultz, P.G., Exploiting chemical libraries, structure, and genomics in the search for kinase inhibitors, *Science*, 281, 530–538, 1998.

15. Heller, R.A., Schena, M., Chai, A., Shalon, D., Bedilion, T., Gilmore, J., Woolley, D.E., and Davis, R.W., Discovery and analysis of inflammatory disease-related genes using cDNA microarrays, *Proc. Natl. Acad. Sci. U.S.A.*, 94, 2150–2155, 1997.
16. Hoffman, P., Grinstein, G., Marx, K., and Grosse, I., DNA visual and analytic data mining, *Proc. 1997 IEEE Visualization Conf.*, Computer Society Press, October 1997, 437–441.
17. Iyer, V.R., Eisen, M.B., Ross, D.T., Schuler, G., Moore, T., Lee, J.C.F., Trent, J.M., Staudt, L.M., Hudson, J., Jr., Boguski, M.S., Lashkari, D., Shalon, D., Botstein, D., and Brown, P.O., The transcriptional program in the response of human fibroblasts to serum, *Science*, 283(5398), 83–87, 1999.
18. Martinetz, T.M., Berkovich, S.G., and Schulten, K.J., Neural-gas network for vector quantization and its application to time-series prediction, *IEEE Trans. Neural Networks*, 4, 558–569, 1993.
19. Marton, M.J., DeRisi, J.L., Bennett, H.A., Iyer, V.R., Meyer, M.R., Roberts, C.J., Stoughton, R., Burchard, J., Slade, D., Dai, H., Bassett, D.E., Hartwell, L.H., Brown, P.O., and Friend, S.H., Drug target validation and identification of secondary drug target effects using DNA microarrays, *Nat. Med.*, 4, 1293-1301, 1998.
20. Michaels, G.S., Carr, D.B., Askenazi, M., Fuhrman, S., Wen, X., and Somogyi, R., Cluster analysis and data visualization of large-scale gene expression data, *Pacific Symp. Biocomputing*, 3, 42–53, 1998.
21. Perou, C.M., Jeffrey, S.S., Rijn, M.V.D., Rees, C.A., Eisen, M.B., Ross, D.T., Pergamenschikov, A., Williams, C.F., Zhu, S. X., Lee, J.C.F., Lashkari, D., Shalon, D., Brown, P.O., and Botstein, D., Distinctive gene expression patterns in human mammary epithelial cells and breast cancers, *Proc. Natl. Acad. Sci. U.S.A.*, 96(16), 9212–9217, 1999.
22. Pietu, G., Alibert, O., Guichard, V., Lamy, B., Bois, F., Leroy, E., Mariage-Sampson, R., Houlgatte, R., Soularue, P., and Auffray, C., Novel gene transcripts preferentially expressed in human muscles revealed by quantitative hybridization of a high density cDNA array, *Genome Res.*, 6, 492–503, 1996.
23. Pollack, J.R., Perou, C.M., Alizadeh, A.A., Eisen, M.B., Pergamenschikov, A., Williams, C.F., Jeffrey, S.S., Botstein, D., and Brown, P.O., Genome-wide analysis of DNA copy-number changes using cDNA microarrays, *Nat. Genet.*, 23(1), 41-46, 1999.
24. Ross, D.T., Scherf, U., Eisen, M.B., Perou, C.M., Rees, C., Spellman, P., Iyer, V., Jeffrey, S.S., Rijn, M.V.D., Waltham, M., Pergamenschikov, A., Lee, J.C.F., Lashkari, D., Shalon, D., Myers, T.G., Weinstein, J.N., Botstein, D., and Brown, P.O., Systematic variation in gene expression patterns in human cancer cell lines, *Nat. Genet.*, 24, 227–235, 2000.
25. SAS Technical Report P-256, Cary, NC: SAS Institute, Inc., 2001.
26. Schena, M., Shalon, D., Heller, R., Chai, A., Brown, P.O., and Davis, R.W., Parallel human genome analysis: microarray-based expression monitoring of 1000 genes, *Proc. Natl. Acad. Sci. U.S.A.*, 93, 10610–10619, 1996.
27. Spellman, P.T., Sherlock, G., Zhang, M.Q., Iyer, V.R., Anders, K., Eisen, M.B., Brown, P.O., Botstein, D., and Futcher, B., Comprehensive identification of cell cycle-regulated genes of the yeast *Saccharomyces cerevisiae* by microarray hybridization, *Molec. Biol. Cell*, 9(12), 3273–3297, 1998.
28. CAMDA: Critical Assessment of Microarray Data Analysis, <http://www.bioinformatics.duke.edu/camda/>, 2000.
29. Compaq biosciences, <http://www.compaq.com/hpc/bio_index.html>.
30. IBM Life Sciences, <http://www-3.ibm.com/solutions/lifesciences/index.html>, 2001.

31. Juno Virtual Supercomputer Network, Juno Online Services, <http://www.juno.com/>, 2001.
32. Parabon Computing, <http://www.parabon.com/>, 2001.
33. Nutec Sciences, <http://www.nutecsciences.com/>, 2001.
34. National Institutes of Health, Genomics and Bioinformatics Group, Cluster Image Map Software, <http://discover.nci.nih.gov/nature2000/tools/cluster.html>, 2001.
35. Scherf, U., Ross, D.T., Waltham, M., Smith, L.H., Lee, J.K., Tanabe, L., Kohn, K.W., Reinhold, W.C., Myers, T.G., Andrews, D.T., Scudiero, D.A., Eisen, M.B., Sausville, E.A., Pommier, Y., Botstein, D., Brown, P.O., and Weinstein, J.N., A gene expression database for the molecular pharmacology of cancer, *Nat. Genet.*, 24, 236–244, 2000; and <http://discover.nci.nih.gov/nature2000/tools/cluster.html>.
36. Breunig Markus, M., Kriegel, H.-P., Ng, R.T., and Sander, J., LOF: identifying density-based local outliers, *SIGMOD Conf. 2000*, 93–104, 2000.
37. Knorr, E.M. and Ng, R.T., Algorithms for mining distance-based outliers in large datasets, *VLDB J.*, 392–403, 1998.
38. Knorr, E.M., Ng, R.T., and Tucakov, V., Distance-based outliers: algorithms and applications, *VLDB J.*, 8(3–4), 237–253, 2000.
39. Saksena, M., France, R.B., and Larrondo-Petrie, M.M., A characterization of aggregation, *Proc. 5th Int. Conf. Object-Oriented Inf. Syst. (OOIS'98)*, 1998.
40. Nagao, K. and Hasida, K., Automatic text summarization based on the global document annotation, Technical Report, Sony Computer Science Laboratory, 1998.
41. Ester, M., Frommelt, A., Kriegel, H.-P., and Sander, J., Algorithms for characterization and trend detection in spatial databases, *Proc. 4th Int. Conf. Knowledge Discovery and Data Mining (KDD'98)*, 1998.
42. Danuser, G. and Stricker, M., Parametric model fitting: from inlier characterization to outlier detection, *IEEE Trans. Pattern Analysis and Machine Intelligence*, 20(2), 263–280, 1988.
43. Vellaikal, A. and Kuo, C.-C. J., Hierarchical clustering techniques for image database organization and summarization, in C.-C.J. Kuo, S.-F. Chang, and S. Panchanathan, Eds., *Multimedia Storage and Archiving Systems III* (VV02), Vol. 3527 of *SPIE Proc.*, SPIE, Boston, MA, 1998, 68–79.
44. Sridhar, R., Rastogi, R., and Shim, K., Efficient algorithms for mining outliers from large data sets, *SIGMOD Conf. 2000*, 427–438, 2000.
45. Agrawal, R., Imielinski, T., and Swami, A., Mining association rules between sets of items in large databases, in *Proc. ACM SIGMOD Conf. Manage. Data*, Washington, D.C., 1993, 207–216.
46. Agrawal, R., Mannila, H., Srikant, R., Toivonen, H., and Verkamo, A.I., Fast discovery of association rules, in *Advances in Knowledge Discovery and Data Mining*, U. Fayyad, G. Piatetsky-Shapiro, P. Smyth, and R. Uthurusamy, Eds., AAAI Press, Menlo Park, CA, 1996.
47. Srikant, R. and Agrawal, R., Mining generalized association rules, *Proc. 21st Int. Conf. Very Large Databases*, Zurich, Switzerland, September 1995.
48. Srikant, R. and Agrawal, R., Mining quantitative association rules in large relational tables, *Proc. ACM SIGMOD Conf. Manage. Data*, Montreal, Canada, 1996.
49. Han, J. and Fu, Y., Discovery of multiple-level association rules from large databases, in *Proc. 21st Int. Conf. Very Large Databases*, Zurich, Switzerland, September 1995.
50. Brin, S., Motwani, R., and Silverstein, C., Beyond market baskets: generalizing association rules to correlations, *SIGMOD Rec. (ACM Special Interest Group on Management of Data)*, 26(2), 265, 1997.

51. Pasquier, N., Bastide, Y., Taouil, R., and Lakhal, L., Discovering frequent closed item sets for association rules, *7th Int. Conf. Database Theory*, 1999.
52. Lu, H., Han, J., and Feng, L., Stockmovement and n-dimensional inter-transaction association rules, *Proc. 1998 SIGMOD Workshop on Research Issues on Data Mining and Knowledge Discovery (DMKD'98)*, Seattle, WA, 1998, 12:1–12:7.
53. Sarda, N.L. and Srinivas, N.V., An adaptive algorithm for incremental mining of association rules, *Proc. DEXA Workshop '98*, 1998, 240–245.
54. Kuok, C.M., Fu, A., and Wong, M.H., Fuzzy association rules in large databases with quantitative attributes, in *ACM SIGMOD Rec.*, 1998.
55. Dong, G. and Li, J., Interestingness of discovered association rules in terms of neighborhood-based unexpectedness, in X. Wu, R. Kotagiri, and K. Korb, Eds., *Proc. Second Pacific-Asia Conf. Knowledge Discovery and Data Mining (PAKDD'98)*, Melbourne, Australia, 1998, 72–86.
56. Megiddo, M. and Srikant, R., Discovering predictive association rules, *Knowledge Discovery and Data Mining*, ACM Press, 1998.
57. Solomon, H., Classification procedures based on dichotomous response vector, *Studies in Item Analysis and Prediction*, H. Solomon, Ed., Stanford University Press, Stanford, CA, 1961, 177–186.
58. Bock, H.H., *Automatic Classification*, Vandenhoeck and Ruprecht, Göttingen, 1974.
59. Thode, H.C., Jr., Mendell, N.R., and Finch, S.J., Simulated percentage points for the null distribution of the likelihood ratio test for a mixture of two normals, *Biometrics*, 44, 1195–1201, 1988.
60. Goryachev, A.B., Macgregor, P.F., and Edwards, A.M., Unfolding of microarray data, *J. Comput. Biol.*, 8(4), 443–461, 2001.
61. Mezzich, J.E. and Solomon, H., *Taxonomy and Behavioral Science*, Academic Press, New York, 1980.
62. Weiss, S.M. and Kulikowski, C.A., *Computer Systems that Learn. Classification and Prediction Methods from Statistics, Neural Nets, Machine Learning, and Expert Systems*, Morgan-Kaufmann, San Francisco, 1991.
63. Young, R., Moore, M., and Moore, J.D., *Computer Systems that Learn: Classification and Prediction Methods from Statistics, Neural Nets, Machine Learning, and Expert Systems*, Morgan-Kaufmann, San Francisco, 1994.
64. Nakhaeizadeh, G., Project STATLOG, Tech. Rep., ESPRIT IPSS-2 Number 5170: Comparative Testing and Evaluation of Statistical and Logical Learning Algorithms for Large-Scale Applications in Classification, Prediction and Control, 1993.
65. Grassmann, J., Statistical classification methods for protein fold class prediction, in A. Prat, Ed., *COMPSTAT. Proc. Computational Statistics. 12th Symp.*, Barcelona, Spain, Physika-Verlag, Heidelberg, 1996, 277–282.
66. Andernach, J.A., A machine learning approach to the classification and prediction of dialogue utterances, in *Proc. Second Int. Conf. New Methods in Language Processing*, 98–109, 1996.
67. Mugica, F. and Nebot, A., A specialization of the k-nearest neighbor classification rule for the prediction of dynamical systems using FIR, in *Advances in Artificial Intelligence and Engineering Cybernetics*, Vol. III, 1996, 130–136.
68. Cuadras, C.M., Fortiana, J., and Oliva, F., Representation of statistical structures, classification and prediction using multidimensional scaling, in W. Gaul and D. Pfeifer, Eds., *From Data to Knowledge*, Springer-Verlag, Berlin, 1996, 20–31.
69. Slawinski, T.Y., Data-based generation of fuzzy-rules for classification, prediction and control with the Fuzzy-ROSA method, *European Control Conf.*, Karlsruhe, Aug./Sept. 1999, VDI/VDE Gesellschaft Mess- und Automatisierungstechnik, 1999.

70. Lim, T.-J., Loh, W.-Y., and Shih, Y.-S., A comparison of prediction accuracy, complexity, and training time of thirty-three old and new classification algorithms, *Machine Learning*, 40(3), 203–228, 1999.
71. Yee, P.V., Regularized Radial Basis Function Networks: Theory and Applications to Probability Estimation, Classification, and Time Series Prediction. Ph.D. thesis, Dept. of ECE, McMaster University, Hamilton, Canada, 1998.
72. Snijders, T. and Nowicki, K., Estimation and prediction for stochastic blockmodels for graphs with latent block structure, *J. Classification*, 14, 75–100, 1997.
73. Lim, T. and Loh, W., A Comparison of Prediction Accuracy, Complexity, and Training Time of Thirty-Three Old and New Classification Algorithms, Technical Report, Department of Statistics, University of Wisconsin-Madison, No. 979, 1997.
74. Driesen, K. and Holzle, U., Improving Indirect Branch Prediction with Source- and Arity-based Classification and Cascaded Prediction, Technical Report TRCS98-07, University of California, Santa Barbara, 1998.
75. Shanahan, J.G., Cartesian Granule Features: Knowledge Discovery of Additive Models for Classification and Prediction, Ph.D. thesis, Dept. of Engineering Maths, University of Bristol, Bristol, U.K., 1998.
76. Tibshirani, R., Bias, Variance and Prediction Error for Classification Rules, Technical Report, University of Toronto, November 1996.
77. Agrawal, R., Gehrke, J., Gunopulos, D., and Raghavan, P., Automatic subspace clustering of high dimensional data for data mining applications, *Proc. ACM SIGMOD Int. Conf. Manage. Data*, 1998, 94–105.
78. Aggarwal, C.C., Procopiuc, C.M., Wolf, J.L., Yu, P.S., and Park, J.S., Fast algorithms for projected clustering, *SIGMOD Conf.*, 1999, 61–72.
79. Aggarwal, C.C. and Yu, P.S., Finding generalized projected clusters in high dimensional spaces, *SIGMOD Conf.*, 2000, 70–81.
80. Anderberg, M.R., *Cluster Analysis for Applications*, Academic Press, New York, 1973.
81. Art, D., Gnanadesikan, R., and Kettenring, R., Data-based metrics for cluster analysis, *Utilitas Mathematica*, 21A, 75–99, 1982.
82. Binder, D.A., Approximations to Bayesian cluster analysis, *Biometrika*, 68, 275–285, 1981.
83. Venkatesh, G., Gehrke, J., and Ramakrishnan, R., CACTUS: clustering categorical data using summaries, *Proc. Int. Conf. Knowledge Discovery and Data Mining*, 1999, 73–83.
84. Sudipto, G., Rastogi, R., and Shim, K., CURE: an efficient clustering algorithm for large databases, *Proc. ACM SIGMOD Int. Conf. Manage. Data*, ACM Press, Seattle, Washington, 1998, 73–84.
85. Sudipto, G., Rastogi, R., and Shim, K., ROCK: a robust clustering algorithm for categorical attributes, *Proc. 15th Int. Conf. Data Engineering*, 23–26 March 1999, 512–521.
86. Spath, H., *Cluster Analysis Algorithms*, Ellis Horwood, Chichester, 1980.
87. Schnell, P., A method for discovering data-groups, *Biometrica*, 6, 47–48, 1964.
88. Englemann, L. and Hartigan, J.A., Percentage points of a test for clusters, *J. Am. Statistical Assoc.*, 64, 1647–1648, 1969.
89. Calinski, T. and Harabasz, J., A dendrite method for cluster analysis, *Commun. Statistics*, 3, 1–27, 1974.
90. Duran, B.S. and Odell, P.L., *Cluster Analysis*, Springer-Verlag, New York, 1974.
91. Hartigan, J.A., *Clustering Algorithms*, John Wiley & Sons, New York, 1975.

92. Hartigan, J.A., Distribution problems in clustering, in *Classification and Clustering*, J. Van Ryzin, Ed., Academic Press, New York, 1977.
93. Hartigan, J.A., Asymptotic distributions for clustering criteria, *Ann. Statistics*, 6, 117–131, 1978.
94. Hartigan, J.A., Consistency of single linkage for high-density clusters, *J. Am. Statistical Assoc.*, 76, 388–394, 1981.
95. Hawkins, D.M., Muller, M.W., and ten Krooden, J.A.,Cluster Analysis, in *Topics in Applied Multivariate Analysis*, D. M. Hawkins, Ed., Cambridge University Press, Cambridge, 1982.
96. Hubert, L.J. and Baker, F.B., An empirical comparison of baseline models for goodness-of-fit in r-diameter hierarchical clustering, in *Classification and Clustering*, J. Van Ryzin, Ed., Academic Press, New York, 1977.
97. Huizinga, D.H., A Natural or Mode Seeking Cluster Analysis Algorithm, Technical Report 78-1, 1978, Behavioral Research Institute, 2305 Canyon Blvd., Boulder, CO, 80302.
98. Hubert, L., Approximate evaluation techniques for the single-link and complete-link hierarchical clustering procedures, *J. Am. Statistical Assoc.*, 69, 698–704, 1974.
99. Hathaway, R.J., A constrained formulation of maximum-likelihood estimation for normal mixture distributions, *Ann. Statistics*, 13, 795–800, 1985.
100. Everitt, B.S., Unresolved problems in cluster analysis, *Biometrics*, 35, 169–181, 1979.
101. Barnett, V., Ed., *Interpreting Multivariate Data*, John Wiley & Sons, New York, 1981.
102. Blashfield, R.K. and Aldenderfer, M.S., The literature on cluster analysis, *Multivariate Behav. Res.*, 13, 271–295, 1978.
103. Koontz, W.L.G. and Fukunaga, K., A nonparametric valley-seeking technique for cluster analysis, *IEEE Trans. Computers*, C-21, 171–178, 1972 .
104. Koontz, W.L.G. and Fukunaga, K., Asymptotic analysis of a nonparametric clustering technique, *IEEE Trans. Computers*, C-21, 967–974, 1972.
105. Koontz, W.L.G., Narendra, P.M., and Fukunaga, K., A graph-theoretic approach to nonparametric cluster analysis, *IEEE Trans. Computers*, C-25, 936–944, 1976.
106. Fukunaga, K., *Introduction to Statistical Pattern Recognition*, Academic Press, San Diego, CA, 1990.
107. Lee, K.L., Multivariate Tests for Clusters, *J. Am. Statistical Assoc.*, 74, 708–714, 1979.
108. MacQueen, J.B., Some methods for classification and analysis of multivariate observations, *Proc. Fifth Berkeley Symposium on Mathematical Statistics and Probability*, 1, 281–297, 1967.
109. Marriott, F.H.C., Practical problems in a method of cluster analysis, *Biometrics*, 27, 501–514, 1971.
110. Marriott, F.H.C., Separating mixtures of normal distributions, *Biometrics*, 31, 767–769, 1975.
111. McClain, J.O. and Rao, V.R., CLUSTISZ: a program to test for the quality of clustering of a set of objects, *J. Marketing Res.*, 12, 456–460, 1975.
112. Milligan, G.W., An examination of the effect of six types of error perturbation on fifteen clustering algorithms, *Psychometrika*, 45, 325–342, 1980.
113. Milligan, G.W., A review of Monte Carlo tests of cluster analysis, *Multivariate Behavioral Res.*, 16, 379–407, 1981.
114. Milligan, G.W. and Cooper, M.C., An examination of procedures for determining the number of clusters in a data set, *Psychometrika*, 50, 159–179, 1985.
115. Mosteller, F. and Tukey, F.W., *Data Analysis and Regression — A Second Course in Statistics*, Addison-Wesley, The Philippines, 1977.
116. Pollard, D., Strong consistency of k-means clustering, *Ann. Statistics*, 9, 135–140, 1981.

117. Sarle, W.S., Cluster analysis by least squares, *Proc. Seventh Annu. SAS Users Group Int. Conf.*, 1982, 651–653.
118. Scott, A.J. and Symons, M.J., Clustering methods based on likelihood ratio criteria, *Biometrics*, 27, 387–397, 1971.
119. Scott, D.W., *Multivariate Density Estimation*, John Wiley & Sons, New York, 1992.
120. Sneath, P.H.A. and Sokal, R.R., *Numerical Taxonomy*, W.H. Freeman, San Francisco, 1973.
121. Symons, M.J., Clustering criteria and multivariate normal mixtures, *Biometrics*, 37, 35–43, 1981.
122. Titterington, D.M., Smith, A.F.M., and Makov, U.E., *Statistical Analysis of Finite Mixture Distributions*, John Wiley & Sons, New York, 1985.
123. Vuong, Q.H., Likelihood ratio tests for model selection and non-nested hypotheses, *Econometrica*, 57, 307–333, 1989.
124. Wolfe, J.H., Pattern clustering by multivariate mixture analysis, *Multivariate Behavioral Res.*, 5, 329–350, 1970.
125. Wong, M.A., A hybrid clustering method for identifying high-density clusters, *J. Am. Statistical Assoc.*, 77, 841–847, 1982.
126. Wong, M.A. and Schaack, C., Using the k-th nearest neighbor clustering procedure to determine the number of subpopulations, *American Statistical Association 1982 Proc. Statistical Computing Sect.*, 1982, 40–48.
127. Berchtold, S. and Keim, D.A., High-dimensional index structures, database support for next decade's applications, *ACM SIGMOD Int. Conf. Manage. Data: 501*, see also <http://www.informatik.uni-trier.de/~ley/db/indices/a-tree/b/Berchtold:Stefan.html>, 1998.
128. Berchtold, S., Böhm, C., and Kriegel, H.-P., The pyramid-technique: towards breaking the curse of dimensionality, *Proc. ACM SIGMOD Int. Conf. Manage. Data*, 142–153, 1998.
129. Ester, M., Kriegel, H.-P., and Xu, X., A database interface for clustering in large spatial databases, *Proc. First Int. Conf. Knowledge Discovery and Data Mining, KDD95*, U.M. Fayyad and R. Uthurusamy, Eds., AAAI Press, ISBN 0-929280-82-2, 1995.
130. Ester, M., Kriegel, H.-P., and Xu, X., *Knowledge Discovery in Large Spatial Databases: Focusing Techniques for Efficient Class Identification, Lecture Notes in Computer Science*, Springer-Verlag, New York, 1995.
131. Ester, M. and Wittmann, R., Incremental generalization for mining in a data warehousing environment, *Proc. Int. Conf. Extending Database Technology*, 1998, 135–149.
132. Chen, M.-S., Han, J., and Yu,V., Data mining: an overview from a database perspective, *IEEE: Transactions on Knowledge and Data Engineering*, 8(6), 866–883, 1996.
133. Hinneburg, A., Keim, D.A., An efficient approach to clustering in large multimedia databases with noise, *Proc. 4th Int. Conf. Knowledge Discovery and Data Mining*, 1998.
134. Hinneburg, A. and Keim, D.A., Optimal grid-clustering: towards breaking the curse of dimensionality in high-dimensional clustering, *Proc. 25th Int. Conf. Very Large Data Bases*, Edinburgh, 1999, 506–517.
135. Hinneburg, A., Keim, D.A., and Wawryniuk, M., HD-eye: visual mining high-dimensional data, *IEEE Computer Graphics and Applications*, 19(5), 22–31, 1999.
136. Hinneburg, A., Aggarwal, C., and Keim, D.A., What is the nearest neighbor in high dimensional spaces, in *Proc. 26th Int. Conf. Very Large Data Bases*, Cairo, 2000.

137. Thomas, J.G., Olson, J.M., Tapscott, S.J., and Zhao, L.P., An efficient and robust statistical modeling approach to discover differentially expressed genes using genomic expression profiles, *Genome Res.*, 11(7), 1227–1236, 2001.
138. Sheikholeslami, G., Chatterjee, S., and Zhang, A., WaveCluster: a multi-resolution clustering approach for very large spatial databases. *Proc. 24th Int. Conf. Very Large Data Bases*, A. Gupta, O. Shmueli, and J. Widom, Eds., Morgan-Kaufmann, San Francisco, 1998.
139. Keim, D.A., Databases and Visualization, Tutorial on ACM SIGMOD Int. Conf. Manage. Data, 1996.
140. Hoffman, P. and Grinstein, G., Multidimensional information visualizations for data mining with applications for machine learning classifiers, in *Information Visualization in Data Mining and Knowledge Discovery*, Morgan-Kaufmann, San Francisco, 2000.
141. Inselberg, A., Grinstein, G., Buja, A., and Asimov, A., Visualizing multidimensional (multivariate) data and relations: perception vs. geometry, *Proc. 1995 IEEE Visualization Conf.*, Nielson and Silver, Eds., 1995, 405–411.
142. Day, W.H. and Edelsbrunner, H., Efficient algorithms for agglomerative hierarchical clustering methods, *J. Classification*, 1(1), 7–24, 1984.
143. Mizoguchi, R. and Shimura, M., A nonparametric algorithm for detecting clusters using hierarchical structure, *IEEE Trans. Pattern Analysis and Machine Intelligence*, PAMI-2, 292–300, 1980.
144. Murtagh, F., Complexities of hierarchic clustering algorithms: state of the art, *Computational Statistics Quart.*, 1, 101–113, 1984.
145. Schikuta, E., Grid clustering: an efficient hierarchical method for very large data sets, *Proc. 13th Conf. Pattern Recognition*, Vol. 2, IEEE Computer Society Press, 1996, 101–105.
146. Ward, J.H., Hierarchical grouping to optimize an objective function, *J. Am. Statistical Assoc.*, 58, 236–244, 1963.
147. Kohonen, T., Self-organized formation of topologically correct feature maps, *Biological Cybernetics*, 43, 59–69, 1982.
148. Kohonen, T., *Self-Organization and Associative Memory*, Springer-Verlag, Berlin, 1989.
149. Kohonen, T., *Self-Organizing Maps*, Springer-Verlag, Berlin, 1995.
150. Timmis, J., Neal, H., and Hunt, J., Data analysis with artificial immune systems and cluster analysis and kohonen networks: some comparisons, *Proc. Int. Conf. Systems and Man and Cybernetics*, IEEE, 1999, 922–927.
151. Banfield, J.D. and Raftery, A.E., Model-based Gaussian and non-Gaussian clustering, *Biometrics*, 49, 803–821, 1993.
152. McLachlan, G.J. and Basford, K.E., *Mixture Models*, Marcel Dekker, New York, 1988.
153. Priebe, C.E., Adaptive mixtures, *J. Am. Statistical Assoc.*, 89, 796–806, 1994.
154. Cooper, M.C. and Milligan, G.W., The effect of error on determining the number of clusters, *Proc. Int. Workshop on Data Analysis, Decision Support and Expert Knowledge Representation in Marketing and Related Areas of Research*, 319–328, 1988.
155. Everitt, B.S., A Monte Carlo investigation of the likelihood ratio test for the number of components in a mixture of normal distributions, *Multivariate Behav. Res.*, 16, 171–180, 1981.
156. Everitt, B.S. and Hand, D.J., *Finite Mixture Distributions*, Chapman & Hall, New York, 1981.
157. Bock, H.H., On some significance tests in cluster analysis, *J. Classification*, 2, 77–108, 1985.

158. Ling, R.F., A probability theory of cluster analysis, *J. Am. Statistical Assoc.*, 68, 159–169, 1973.
159. Lindsay, B.G. and Basak, P., Multivariate normal mixtures: a fast consistent method of moments, *J. Am. Statistical Assoc.*, 88, 468–476, 1993.
160. Lauritzen, S.L., The EM algorithm for graphical association models with missing data, *Computational Statistics and Data Analysis*, 19, 191–201, 1995.
161. Palmer, C.R. and Faloutsos, C., Density biased sampling: an improved method for data mining and clustering, *SIGMOD2000, Int. Conf. Manage. Data*, May 2000, Dallas, TX, 2000, 82–92.
162. Bezdek, J.C., *Pattern Recognition with Fuzzy Objective Function Algorithms*, Plenum Press, New York, 1981.
163. Bezdek, J.C. and Pal, S.K., Eds., *Fuzzy Models for Pattern Recognition*, IEEE Press, New York, 1992.
164. Bensmail, H., Celeux, G., Raftery, A.E., and Robert, C.P., Inference in model-based cluster analysis, *Statistics and Computing*, 7, 1–10, 1997.
165. Fritzke, B., A growing neural gas network learns topologies, in G. Tesauro, D.S. Touretzky, and T.K. Leen, Eds., *Advances in Neural Information Processing Systems 7*, MIT Press, Cambridge, MA, 1995.
166. Fritzke, B., The LBG-U method for vector quantization — an improvement over LBG inspired from neural networks, *Neural Processing Lett.*, 5(1), 35–45, 1997.
167. Gersho, A. and Gray, R.M., *Vector Quantization and Signal Compression*, Kluwer Academic, Boston, 1992.
168. David, G., Kleinberg, J.M., and Raghavan, P., Clustering categorical data: an approach based on dynamical systems, *Proc. 24th Int. Conf. Very Large Data Bases*, A. Gupta, O. Shmueli, and J. Widom, Eds., Morgan Kaufmann, San Francisco, 1998, 311–322.
169. David, G., Kleinberg, J.M., and Raghavan, P., Clustering categorical data: an approach based on dynamical systems, *VLDB J.*, 8, 222–236, 2000.
170. Kaufmann, L. and Rousseeuw, P.J., *Finding Groups in Data: An Introduction to Cluster Analysis*, John Wiley & Sons, New York, 1990.
171. Kearns, M., Mansour, Y., and Ng, A., An information-theoretic analysis of hard and soft assignment methods for clustering, *Proc. 13th Conf. Uncertainty in Artificial Intelligence*, Morgan-Kaufmann, San Francisco, 1997, 282–293.
172. Rojas, R., *Neural Networks — A Systematic Introduction*, Springer-Verlag, Berlin, 1996.
173. Tukey, J.W., Mathematics and the picturing of data, *Proc. Int. Congr. Mathematicians*, Vancouver, Canada, 2, 523–531, 1975.
174. Tukey, J.W., *Exploratory Data Analysis*, Addison-Wesley, Reading, MA, 1977.
175. Friedman, J.H. and Tukey, J.W., A projection pursuit algorithm for exploratory data analysis, *IEEE Trans. Computing*, C-23, 881–889, 1974.
176. Becker, R.A., Cleveland, W.S., and Wilks, A.R., Dynamic graphics for data analysis, *Statistical Sci.*, 2, 355–395, 1987.
177. Becker, R.A. and Cleveland, W.S., Brushing scatterplots, *Technometrics*, 29, 127–142, 1987; also in *Dynamic Graphics for Statistics*, W.S. Cleveland and M.E. McGill, Eds., Wadsworth, Belmont, CA, 1988, 201–224.
178. Chambers, J.M., Cleveland, W.S., Kleiner, B., and Tukey, P.A., *Graphical Methods in Data Analysis*, Wadsworth, Belmont, CA, 1976.
179. Cleveland, W.S. and Devlin, S.J., Locally weighted regression: an approach to regression analysis by local fitting, *J. Am. Statistical Assoc.*, 83, 596–610, 1988.
180. Cleveland, W.S. and McGill, M.E., *Dynamic Graphics for Statistics*, Wadsworth, Belmont, CA, 1988.

181. Cleveland, W.S., Devlin, S.J., and Grosse, E., Regression by local fitting: methods, properties, and computational algorithms. Nonlinear modeling and forecasting, *J. Econometrics*, 37, 87–114, 1988.
182. Cleveland, W., Grosse, E., and Shyu, W., Local regression models, in *Statistical Models*, S.J. Chambers and T. Hastie, Eds., Wadsworth, Belmont, CA, 1992, 309–376.
183. Cleveland, W., *Visualizing Data*, Hobart Press, Summit, NJ, 1993.
184. Cleveland, W., *The Elements of Graphing Data*, Hobart Press, Summit, NJ, 1994.
185. Cleveland, W.S. and Loader, C., Smoothing by Local Regression: Principles and Methods, Technical Report, AT&T Bell Laboratories, Murray Hill, NJ, 1995.
186. Fayyad, U., Grinstein, G., and Wierse, A., *Information Visualization in Data Mining and Knowledge Discovery*, Morgan-Kaufmann Publishers, San Francisco, 2001.
187. Grinstein, G. and Levkowitz, H., Eds., *Perceptual Issues in Visualization*, Springer-Verlag, Berlin, 1995 .
188. Grinstein, G., Wierse, A., and Lang, U., Eds., *Proc. Second IEEE Workshop on Issues on the Integration of Databases and Visualization, Lecture Notes in Computer Science*, Vol. 1183, Springer-Verlag, Berlin, 1996.
189. Grinstein, G., Human interaction in database and visualization integration, *Proc. 1995 IEEE Visualization Second Workshop on Issues on the Integration of Databases and Visualization, Lecture Notes in Computer Science*, Vol. 1183, Springer-Verlag, Berlin 1996.
190. Grinstein, G., Visualization and data mining, *Proc. 1996 Int. Conf. Knowledge Discovery in Databases*, August 1996, Portland, 1996, 384–385.
191. Grinstein, G. and Meneses, C., Visual data exploration in massive data sets, in *Information Visualization in Data Mining and Knowledge Discovery*, Morgan-Kaufmann, San Francisco, 2001.
192. Laskowski, S. and Grinstein, G., Requirements for benchmarking the integration of visualization and data mining, in *Information Visualization in Data Mining and Knowledge Discovery*, Morgan-Kaufmann, 2000.
193. Lee, J.P. and Grinstein, G., Eds., *Proc. IEEE Workshop on Issues on the Integration of Databases and Visualization, Lecture Notes in Computer Science*, Vol. 871, Springer-Verlag, Berlin, 1994.
194. Lee, J.P. and Grinstein, G., An Architecture for Retaining and Analyzing Visual Explorations of Databases, *Proc. 1995 IEEE Visualization Conference*, Nielson and Silver, Eds., 1995, 101–108.
195. Lee, J. and Grinstein, G., Describing Visual Interactions to the Database: Closing the Loop Between Users and Data, *Proc. 1996 SPIE Visual Data Exploration and Analysis Conference*, San Jose, CA, 2656, 93–103, 1996.
196. Meneses, C. and Grinstein, G., Visualization for enhancing the data mining process, *Proc. SPIE, Data Mining and Knowledge Discovery: Theory, Tools, and Technology III*, Vol. 4384, 2001.
197. Unwin, A.R., Hawkins, G., Hoffman, H., and Siegl, B., Interactive graphics for data sets with missing values — MANET, *J. Computational and Graphical Statistics*, 5(2), 113–122, 1996.
198. Swayne, D.F. and Buja, A., Missing Data in Interactive High-Dimensional Visualization, *Computational Statistics*, 12(1), 1997.
199. Inselberg, A. and Dimsdale, B., Parallel coordinates: a tool for visualizing multidimensional geometry, *IEEE Visualization '90*, San Francisco, CA, 1990, 361–370.
200. Lachiche, N. and Marquis, P., Scope classification: an instance-based learning algorithm with a rule-based characterization, in *Proc. Tenth ECML*, Chemnitz, Germany, Springer-Verlag, Berlin, 1998.

7 Data Management in Microarray Fabrication, Image Processing, and Data Mining

Alexander Kuklin, Shishir Shah, Bruce Hoff, and Soheil Shams

CONTENTS

Microarray technology provides an unprecedented means for carrying out high-throughput gene expression analysis experiments. Microarrays are becoming indispensable tools for investigating the mechanism of drug action. Gene expression patterns (or "signatures") can be promptly obtained for cultured cells, and animal or patient samples before, after, or during a drug treatment course by using powerful

0-8493-2285-5/02/$0.00+$1.50

microarray software systems.[1] Gene expression monitoring has been successfully used in molecular classification of cancer and has surpassed in accuracy morphological and enzyme-based histochemical analyses that are traditionally used in clinical cases.[2]

In the maturation process of microarray technology, there are several kinds of challenges. One is to develop the hardware for conducting hybridization experiments. Another challenge is to manage the massive amount of information associated with this technology so that the results can yield insight into the genomic functions in biological systems. With the steady progress of the developments of the hardware technology, currently available equipment can reliably produce an image on the order of 10,000 spots and more on single microscope slides. That is, the fundamental challenge from hardware has been mostly resolved.

On the other hand, the informatics challenge has just started. There are three major issues involved. The first is to keep track of the information generated at the stages of chip production and hybridization experiment. The second is to process microarray images to obtain the quantified gene expression values from the arrays. The third is to mine the information from the gene expression data. This chapter outlines an integrated approach to managing the flow of information from microarray fabrication to mining of expression data.

7.1 INTRODUCTION

The major purpose of building a database system for array informatics is to facilitate the access of information for experimental design, problem tracking, and system tuning, replicating chips and experiments, and data sharing with other processes.[3] A considerable problem with regard to establishing widely used standards has been the diversity of the user base. Many companies and laboratories are developing their own specialized databases and user interfaces, all of which are intended to work with internal projects. However, the lack of integration with modules — such as image processing and gene expression data analysis, as well as the expense involved in setting up an in-house database — poses obstacles to many institutions.

One way to overcome complications associated with accumulating and processing data is to use robust and well-established software modules that communicate with each other and could be used both as a system or as independent tools. Such an approach allows the user to prioritize expenses and build up on achieved results in terms of budgeting.

7.2 MICROARRAY FABRICATION INFORMATICS

The first step in microarray fabrication is to design the microarray. Multiple arrays should be "virtually" designed and visualized before a specific arrayer program is written. A researcher may relocate the spots on the blueprint and arrange the appropriate "housekeeping" genes used in normalization.

One of the challenges in the design of high-density arrays is establishing a mapping between the wells in the PCR plates to the spots on the chip and maintaining a database of information associated with the array fabrication process. A user should be able to get information about gene identity, plate, well location in the plate, spot location on the slide, as well as conditions under which the array was printed. The second goal is to have uniformity in the microarray design approach within an institution/company. A researcher needs the ability to design microarrays and work with a microarray-spotting robot from any manufacturer, and (or) submit the microarray design to a core facility without any misunderstanding of the microarray blueprint or the slide-printing procedure. The information flow in the micorarray fabrication process involves several transformations that cannot be handled by simple flat-file operations (e.g., Excel spreadsheets). A tool that offers querying and archiving for plate numbers (which can be bar-coded), size, and IDs for each clone in each well, as well as information about different arrays produced is CloneTracker™ (BioDiscovery, Los Angeles, CA).

7.2.1 Configuring the Slide Settings and Pin Parameters

The CloneTracker software package allows the user to choose and specify all the parameters for printing the slide and archive the information for future reference (Figure 7.1). The first step involves specifying the slide or membrane settings, that is, the size of the slide (membrane) and the parameters for printing area. A user can specify the number of rows and columns of slides on the arrayer platform. The system then automatically calculates all other optional printing parameters after selection of the above numbers.

The pin configuration module in CloneTracker offers several options, depending on the pinhead setup of the microarray printer. This approach gives flexibility to use an in-house arrayer or any microarraying robot from any manufacturer. After these parameters are selected, the operator needs to identify what type of PCR plates are going to be used in the printing procedure (i.e., 96- or 384-well plates).

7.2.2 Arraying Strategy and Microarray Design

CloneTracker allows flexibility in configuring the printing strategy (Figure 7.1). There are many possibilities for a spot layout, which is defined as the array of spots that each pin will dip on the slide for each plate. For example, if using 96-well plates and the arrayer print head is 2×2, then $96/(2 \times 2) = 24$ times each pin will dip a spot on the slide. The spot layout will range from 1×24, 2×12, 3×8, through 24×1. The user needs to identify the plate arraying direction and the drawing direction from each plate. CloneTracker will use this information and design the array in seconds.

Different color filters can be used to color-code rows, columns, spots from plates, or spots for normalization. A researcher can keep track of slides used in the process, type of printing and printing parameters, hybridization notes, etc. The information archived by CloneTracker is necessary for analysis of the expressed data after image and data processing.

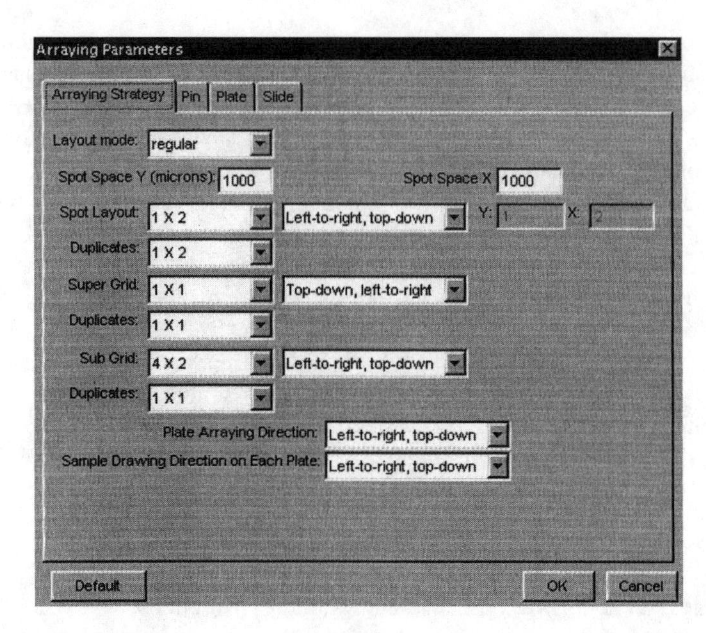

FIGURE 7.1 The Arraying Strategy tab of CloneTracker allows setting the horizontal and vertical spaces between adjacent spots, the spot layout, and spot layout dipping direction and other parameters.

7.2.3 Advantages of Virtual Microarray Design and Data Query

Investment in large databases is initially prohibitive. Additionally, researchers in academia and industry tend to share laboratory hardware, thus necessitating a standardized approach in microarray design. Excel spreadsheets are helpful only at the beginning. Without the means for central storage and management of data, potential for loss of information (e.g., a missing Excel file) increases with the number of arrays being produced. Therefore, the need for an affordable microarray design tool, which can archive and query data, is pertinent (Figure 7.2). CloneTracker has been bundled with several printing robots and in one case has been used as instrument-controlling software (enhanced by Cartesian Technologies, Irvine, CA). The system's output files are useful in tracking experimental data. The acceptance of a consistent technique in microarray design is helpful in the overall process of microarray production and usage in an institution. The design goal of CloneTracker has been to enable researchers without existing database tools to immediately use its built-in SQL database. For those users with existing databases, CloneTracker can directly interface with these databases and become integrated directly into a lab's information management system.

7.3 MICROARRAY IMAGE ANALYSIS

The fundamental goal of array image processing is to measure the intensity of the arrayed spots into quantified expression values. This simple task of reducing an

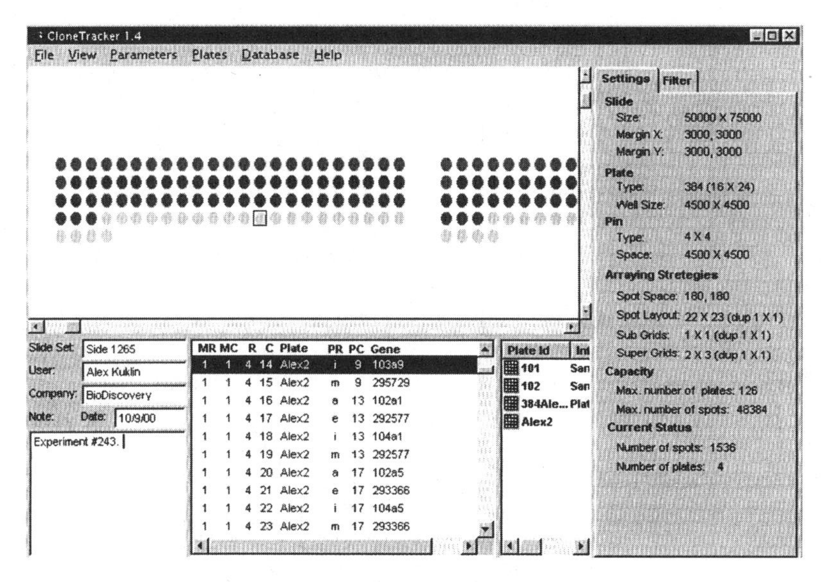

FIGURE 7.2 CloneTracker offers viewing of the microarray before it is printed and querying of data. A click on a spot will bring all the information about the spot location, its plate source, and the location of the plate well. Moreover, color coding of sub-grids makes querying intuitive.

image of spots of varying intensities into a table associating expression values to each spotted gene has been impeded by several microarray technology challenges.

One of the primary concerns is the maximization of the signal-to-noise ratio (SNR) in the raw image. This quantity signifies how well one can resolve a true signal from the noise in the system. The SNR within an image can be estimated by computing the peak signal divided by the variation in the signal. If the scanning system generating the image has a poor SNR, an accurate quantification of individual spots becomes very difficult. We characterize and model all imaging systems as linear and hence noise characterization is implicit in the process. Knowing the functionality of a particular imaging system can help in improving the SNR, and thereby aid in performing an accurate analysis of the signal. Because the image is a function of the incident light or photons, a photomultiplier tube is used to detect the photons and quantize the value to report intensities in an image. Due to thermal emissions and current leakage, electronic noise is introduced in the system. Such noise can typically be modeled as Gaussian-distributed white noise. One can reduce the amount of such noise in the image by controlling exposure time and gain while scanning.

The second noise type introduced in the system is related to number of incident photons. Due to the natural distribution of photons, as the number of incident photons is increased, the variability in its number also increases. Thus, as the intensity increases, the noise also increases. This variability is proportional to the square root of the signal. The third source of noise important for image analysis is independent of the imaging system. The substrate of the slide can have large effects on the SNR. The substrate material used can contribute to background fluorescence, which will reduce the sensitivity of the signal. Improper treatment of the slide surface will result in poor attachment of the DNA, and hence reduce the resultant signal. Finally, residual material as well as improper handling and drying of the slide will result in background fluorescence.

An additional problem in image analysis is spot position variation, which can be caused by inherent mechanical limitations of the spotting process (e.g., vibration of the pins within the pinhead, vibrations of the platform, etc.). Any approach involving even the smallest human effort per spot is certainly impractical for arrays with tens of thousands of spots. Therefore, accurate and powerful software is required to handle this analysis.

Processing of images should not be dependent on the type of scanner or microarray matrices used. Users can work with both glass and high-density membrane microarrays and still be able to process the images with both high efficiency and high accuracy. Such a software tool, designed to drastically reduce the operator's time and effort in microarray data extraction and offer flexibility for the type of microarrays used, is ImaGene™ (BioDiscovery).[4] A user specifies the bounding area or landmarks of the array and the software automatically locates each spot. Graphical user interface (GUI) tools for manual corrections of any possibly misidentified spots are provided to achieve high-accuracy results. ImaGene can analyze image files in many standard formats (e.g., TIFF, GEL, BAS, etc.) from any scanner, thus allowing flexibility to the individual researcher whose experiments are carried out in a core facility or another laboratory.

Additionally, ImaGene has Internet connectivity. Because gene names are imported as a file from CloneTracker, a click of the mouse on any data analysis spot automatically logs on to a user-specified database and finds the information about the gene of interest.

7.3.1 Automation in Image Processing

With the growing number of microarray technology applications in academic research and drug discovery laboratories, it has become necessary to process many microarray images per day and eliminate human involvement in the process. The goals of complete automation in microarray image processing are to provide high accuracy in spot location without human intervention, eliminate noise signals from the data analysis process, and minimize operator involvement in the procedure.

AutoGene™ is an advanced image analysis tool for high-throughput gene expression microarrays and high-density membrane data analysis. The system has been designed to fully automate image analysis and data quantification operations and answer the needs of the pharmaceutical drug discovery laboratories and academic core facilities.

The AutoGene system needs only the input of the microarray configuration (e.g., number of rows and columns of spots) and a list of image files to process, after which analysis is performed automatically. The system will search the image for grid position without the operator's help. It identifies the layout of the array, localizes the spots, and performs measurements without the need for user intervention.

Spots are individually quantified and assessed for artifacts using several patent-pending computer vision algorithms, which increases reliability of the data (Figure 7.3). The software can automatically detect and remove contamination from

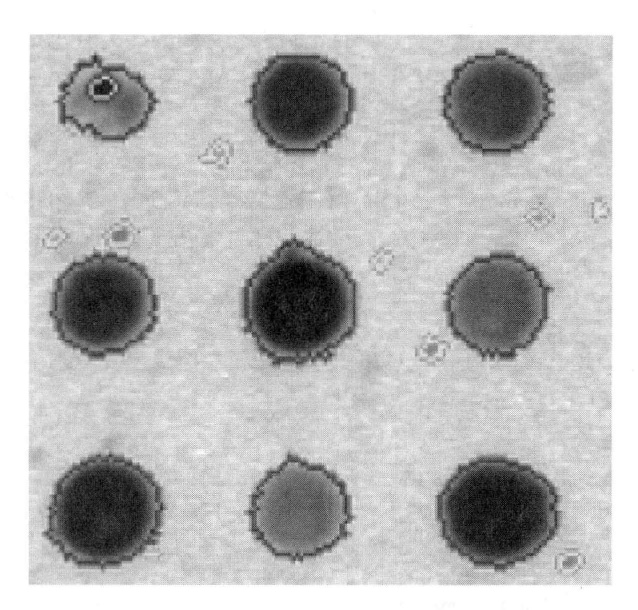

FIGURE 7.3 Advanced patent-pending algorithms of AutoGene compensate for the irregularity of spot sizes and shapes. Areas within a spot, which are filtered out due to preset criteria for signal intensities, will be excluded from the analysis, thus providing high-quality data for data mining.

the images. Robust statistical algorithms process the irregularity of spot sizes and shapes, which is caused by spot-printing hardware errors.

Multiple quality measures per spot are obtained, which allow for association of confidence values to each measurement. The program correctly identifies signal pixels from background, and throws away the contamination pixels. The measurements are of high accuracy, thus providing quality data.

Autonomous operation offers batch-mode (overnight) processing of multiple images. An operator will choose all images to be processed and their computer-drive locations. The system offers consistent quality because results do not vary with the experience of the operator. Human labor cost and training are reduced and throughput is increased.

Visual presentation of the results allows for manual inspection of AutoGene output at any time. Processed data by AutoGene can be accessed through ResultsReviewer™ from different computers connected to the Internet.

AutoGene and ResultsReviewer are equipped with visualization tools to view data in scatterplots, GenePie™, and ratio histograms.

7.4 DATA ANALYSIS AND VISUALIZATION TECHNIQUES

The image processing and analysis step produces a large number of quantified gene expression values. Data typically represent thousands or tens of thousands of gene expression levels across multiple experiments. To make sense of this much data, the use of various visualization and statistical analysis techniques is unavoidable. One of the typical microarray data analysis goals is to find statistically significantly up- or downregulated genes; in other words, outliers or "interestingly" behaving genes in the data.[5] Another possible goal would be to find functional groupings of genes by discovering similarity or dissimilarity among gene expression profiles, or predicting the biochemical and physiological pathways of previously uncharacterized genes.[6] In the following sections, visualization approaches and algorithms implementing various multivariate techniques are discussed that could help realize the above-mentioned goals and provide solutions to the needs of the microarray research community. All these techniques have been implemented in the GeneSight™ software package (BioDiscovery).

7.4.1 Scatterplot

Probably the simplest analysis tool for microarray data visualization is the scatterplot. In a scatterplot, each point represents the expression value of a gene in two experiments: one assigned to the x-axis and the other to the y-axis. In such a plot, genes with equal expression values in both experiments would line up on the identity line (diagonal). Genes that are differentially expressed will be plotted away from the diagonal. The further away a gene is plotted from the identity line, the larger the difference between its expressions in one experiment compared with the other. The absolute strength of the expression levels can be readily visualized in this plot because the higher expression values are plotted further away from the origin.

7.4.2 Principal Component Analysis

It is easy to see why the scatterplot is an ideal tool for comparing the expression profile of genes in two experiments. Three experiments could be plotted and compared in a three-dimensional scatterplot. Three-dimensional plots can be rendered and manipulated on a computer screen. However, when more than three experiments are to be analyzed and compared, the simple scatterplot cannot be used. In the case of 20 experiments, for example, we cannot draw a 20-dimensional plot. Fortunately, there are mathematical techniques available for dimensionality reduction, such as Principal Component Analysis (PCA),[7–9] in which high-dimensional space can be projected down to two or three dimensions (which we can plot). The goal of all dimensionality reduction methods is to perform this operation while preserving most or all the variances of the original dataset. In other words, if two points are relatively "close" to one another in the high-dimensional space, they will be relatively close in the lower-dimensional space as well. In general, it is not possible to perfectly keep this relationship. Imagine a three-dimensional spherical orange peel made flat on the two-dimensional surface of a table. Most of the points lying on the surface of the orange peel keep their relationship to their neighbors. However, we must tear the orange peel in several places, thus breaking the neighboring relationship between some points on the orange peel.

PCA is a method that attempts to preserve the neighboring relationships as much as possible. This multivariate technique is frequently used to provide a compact representation of large amounts of data by finding the axes (principal components) on which the data varies the most. In PCA, the coefficients for the variables are chosen such that the first component explains the maximal amount of variance in the data. The second principal component is perpendicular to the first one and explains the maximum of the residual variance. The third component is perpendicular to the first two and explains the maximum residual variance. This process is continued until all the variance in the data is explained. The linear combination of gene expression levels on the first three principal components can easily be visualized in a three-dimensional plot. This method, just like the scatterplot, provides an easy way of finding outliers in the data; that is, genes that behave differently than most of the genes across a set of experiments. It can also reveal clusters of genes that behave similarly across different experiments. PCA can also be performed on the experiments to find out possible groupings and/or outliers of experiments. In this case, every point plotted in the three-dimensional graph would represent a unique microarray experiment. Points that are placed close to one another represent experiments that have similar expression patterns. Recent findings show that this method should be able to even detect moderate-sized alterations in gene expression.[6] In general, PCA provides a rather practical approach to data reduction, visualization, and identification of unusually behaving outlier genes and/or experiments.

7.4.3 Parallel Coordinate Planes

Two- and three-dimensional scatterplots and PCA plots are ideal for detecting significantly up- or downregulated genes across a set of experiments. These methods, however, do not provide an easy way of visualizing the progression of gene expression over several experiments. These types of questions usually arise in time-series experiments where, for example, gene expression is measured at two-hour intervals. The important question in this case is how gene expression values vary over the duration of the entire experiment. The parallel coordinate planes plotting technique is an ideal visualization tool to answer these types of questions. With this method, experiments are ordered on the x-axis and expression values plotted on the y-axis (Figure 7.4). All genes in a given experiment are plotted at the same location on the x-axis, while their y locations are varied. Another experiment is plotted at another x location in the plane. Typically, the progression of time would be mapped into the x-axis by having higher x values for experiments done at a later time. By connecting the expression values for the same genes in the different experiments, one can obtain a very intuitive way of depicting the progression of gene expression.

7.4.4 Cluster Analysis

Another frequently asked question related to microarrays is finding groups of genes with similar expression profiles across a number of experiments. The most commonly used multivariate technique to find such groups is cluster analysis. Essentially, such techniques arrive at an ordering of the data, which groups (clusters) genes with similar expression patterns closer to each other. These techniques can help establish functional groupings of genes or predict the biochemical and physiological pathways of previously uncharacterized genes.

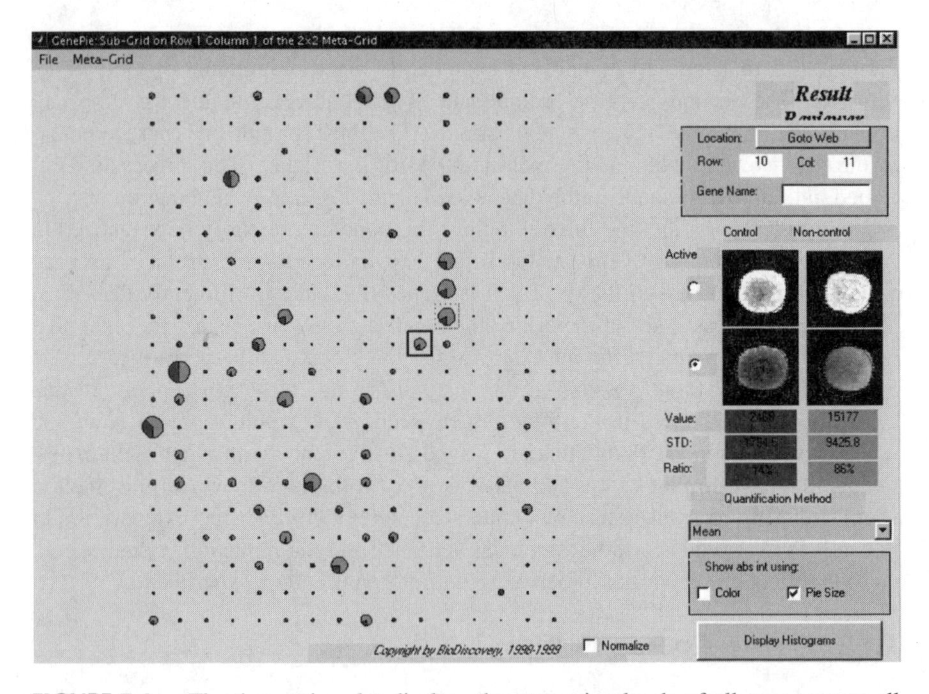

FIGURE 7.4 The time-series plot displays the expression levels of all genes across all experiments/files in the analysis. Experiments are plotted on the x-axis and expression levels of genes across all experiments are plotted on the y-axis. Clicking on a timeline, which finds the corresponding gene, or vice versa, monitors changes of expression levels over time. (See Color Figure 7.4 following page 82.)

The clustering method most frequently found in the literature for finding groups in microarray data is hierarchical clustering.[7,10] This method attempts to group genes and/or experiments in small clusters and then to group these clusters into higher-level clusters, etc. As a result of this clustering or grouping process, a tree of connectivity of observations emerges that can easily be visualized as dendrograms. For gene expression data, not only the grouping of genes but also the grouping of experiments might also be important. When both are considered, it becomes easy to simultaneously search for patterns in gene expression profiles and across many different experimental conditions. For example, a group of genes behaving similarly (e.g., all upregulated) can be seen in a particular group of experiments.

Although hierarchical clustering is currently a commonly employed way of finding groupings in the data, other nonhierarchical (k-means) methods are likely to gain increased popularity in the future with the rapidly growing amounts of data and the ever-increasing average experiment size. A common characteristic of nonhierarchical approaches is to provide sufficient clustering without having to create the full distance or similarity matrix while minimizing the number of scans of the entire dataset.

7.5 DATA NORMALIZATION AND TRANSFORMATION

Several of visualization and statistical techniques described above may be useful for the analysis of microarray data. However, it is important to realize that even with the most powerful statistical methods, the success of the analysis is crucially dependent on the "cleanness" and statistical properties of the data. Essentially, the two questions to ask before even starting any analysis are

1. Does the variation in the data represent the true variation in expression values, or is it contaminated by differences in expression due to experimental variability?
2. Is the data "well-behaving" in terms of meeting the underlying assumptions of the statistical analysis techniques that are applied to it?

It is easy to appreciate the importance of the first question. The significance of the second question derives from the fact that most multivariate analysis techniques are based on underlying assumptions such as normality and homoscedasticity. If these assumptions are not met, at least approximately, then the entire statistical analysis could be distorted and statistical tests might be invalid. Fortunately, there are a variety of statistical techniques available to help us answer "Yes" to the above questions. These are normalization (standardization) and transformation.

Normalization can help us separate true variation in expression values from differences due to experimental variability. This step is necessary because it is quite possible that, due to the complexity of creating, hybridizing, scanning, and quantifying microarrays, variation originating from the experimental process contaminates the data. During a typical microarray experiment, many different variables and parameters can possibly change and hence differentially affect the measured expression levels. Among these are slide quality, pin quality, amount of DNA spotted, accuracy of the arraying device, dye characteristics, scanner quality, and quantification software characteristics, just to name a few. The various methods of normalization aim at removing or at least minimizing expression differences due to variability in any of these types of conditions.

As discussed here, the various transformation methods all aim at changing the variance and distribution properties of the data in such a way that it would be closer in meeting the underlying assumptions of the statistical techniques applied to it in the analysis phase. The most common requirements of statistical techniques are for the data to have homologous variance (homoscedasticity) and normal distribution (normality). In the following, several popular ways of normalizing and transforming microarray data are discussed.

7.5.1 NORMALIZATION OF THE DATA

One of the most popular ways to control for spotted DNA quantity and other slide characteristics is to do a type of local normalization by using two channels (e.g., red and green) in the experiment. For example, a Cy5 (red) labeled probe could be used as control prepared from a mixture of cDNA samples or from normal cDNA. Then a Cy3 (green) labeled experimental probe could be prepared from cDNA extracted from a tumor tissue. The normalized expression values for every gene would then be calculated as the ratio of experimental to control expression. This method can obviously eliminate a large portion of the experimental variation by providing every spot (gene) in the experiment with its own control. Developing on these ideas, three-channel experiments are underway in which one channel serves as the control for the other two. In this case, the expression values of both experimental channels would be divided by the same control value.

In addition to the local normalization method described above, global methods are also available in the form of "control" spots on the slide. Based on a set of these "control" spots, it is possible to control for global variation in overall slide quality or scanning differences.

The above procedures describe some physical measures in terms of spotting characteristics that one can take to normalize the microarray data. However, even after the most careful two- or more-channel spotting with the use of control spots, it is still possible that undesired experimental variation can contaminate the expression data. On the other hand, it is also possible that all or some of these physical normalization techniques are missing from the experiment, in which case it is even more important to find alternative ways of normalization. Fortunately, for both of these scenarios, additional statistics-based normalization methods are available to further clean up the data. As an example, the situation can happen that, for the same set of genes, expression values in one experiment are consistently and significantly different from another experiment due to quality differences between the slides or the printing or scanning process, or possibly due to some other factor (manuscript submitted).

7.5.2 Transformation of the Data

Although there are many different data transforms available, the most frequently used procedure in the microarray literature is to take the logarithm of the quantified expression values.[6,10] An often-cited reason for applying such a transform is to equalize variability in the typically wildly varying raw expression scores. If the expression value was calculated as a ratio of experimental over control conditions, then an additional effect of the log-transform will be to equate up- and downregulation by the same amount in absolute value scores ($\log_{10} 2 = 0.3$ and $\log_{10} 0.5 = -0.3$). Another important side effect of the log-transform is bringing the distribution of the data closer to normal. Having reasonable grounds of meeting the normality and homoscedasticity assumptions after the log-transform, the use of a variety of parametric statistical analysis methods is also much better justified.[8]

7.6 A COMPLETE SOFTWARE SUITE FOR THE MICROARRAY INFORMATION PROCESS

A microarray experiment generates an avalanche of information. Data management is an important aspect of microarray technology. The tools for microarray information analysis described in this chapter (CloneTracker, ImaGene, GeneSight, and AutoGene) have been developed as stand-alone modules. However, they communicate with each other, which allows for access to each step of the information management process.[11,12]

As an example, analysis of several experiments can identify a particularly interested expression pattern of a group of genes involved in a biochemical pathway. If a gene known to be part of this pathway does not show the expected pattern, the researcher can access the processed microarray image and expect a possible operator's error in the image-processing step. Access to spot images and slide-to-PCR plates maps makes data analysis easy and reliable.

7.7 SUMMARY

Although microarray-based analysis and exploration faces technical hurdles, there is room for optimism. There is the potential for unprecedented throughput with a high degree of accuracy. Microarray technology is just in its infancy, and further improvements will ensure that it matures into an even more powerful analytical tool for accurate and high-throughput genomic analysis. In the next decade, microarrays may well become as essential as PCR is now. Common standards will be required before investigators can meaningfully extract and combine data, which will have to be stored in databases that can be effectively mined. Despite the early stage of development of large-scale gene expression monitoring systems and methods, this new technology has already proven exceptionally useful in expanding our knowledge of even well-understood aspects of cellular biology.

From an information processing perspective, microarray technology aids the researcher in transforming and supplementing data available on genes and cells into useful information about gene expression — and ultimately, cellular biology. In this chapter we have described the various issues involved in this process, beginning with the data management and tracking involved in the production of the arrays, to analysis of high-density array images, and finally the analysis of the massive amount of data generated by this technology. However, we must also realize that there remain many challenges in addressing the software needs of the microarray technology. Data quality and the accuracy of data extracted from microarray images form the bases of a well-designed software system.

REFERENCES

1. Debouck, C. and Goodfellow, P., DNA microarrays in drug discovery and development, *Nature Genetics Suppl.*, 21, 48, 1999.
2. Tamayo, P., Slonim, D., Mesirov, J., Zhu, Q., Kitareewan, S., Dmitrowsky, E., Lander, E.S., and Golub, T.R., Interpreting patterns of gene expression with self-organizing maps: methods and application to hematopoietic differentiation, *Proc. Natl. Acad. Sci. U.S.A.*, 96, 2907, 1999.
3. Zhou, Y.-X., Kalocsai, P., Chen, J.-Y., and Shams, S., Information processing issues and solutions associated with microarray technology, *Microarray Biochip Technology*, M. Schena, Ed., Eaton Publishing, MA, 2000, 167–200.

4. Kuklin, A., Using array image analysis to combat HTS bottlenecks, *GEN*, 19, 32, 1999.
5. Heyer, L.J., Kruglyak, S., and Yooseph, S., Exploring expression data: identification and analysis of coexpressed genes, *Genome Res.*, 9, 1106, 1999.
6. Hilsenbeck, S.G., Friedrichs, W.E., Schiff, R., O'Connell, P., Hansen, R.K., Osborne, C.K., and Fuqua, S.A.W., Statistical analysis of array expression data as applied to the problem of Tamoxifen resistance, *J. Natl. Cancer Inst.*, 91, 453, 1999.
7. Duda, R.O. and Hart, P.E., *Pattern Classification and Scene Analysis*, John Wiley & Sons, New York, 1973.
8. Johnson, R.A. and Wichern, D.A., *Applied Multivariate Statistical Analysis*, Prentice-Hall, Englewood Cliffs, NJ, 1998.
9. Raychaudhuri, S., Stuart, J.M., and Altman, R.B., Principal components analysis to summarize microarray experiments: application to sporulation time series, *Proc. Pacific Symp. BioComputing*, 5, 452, 2000.
10. Eisen, M.B., Spellman, P.T., Brown, P.O., and Botstein, D., Cluster analysis and display of genome-wide expression patterns, *Proc. Natl. Acad. Sci. U.S.A.*, 95, 14863, 1998.
11. Kuklin, A., Automation in microarray image processing and data mining, *High Throughput Screening*, January, 4, 2000.
12. Kalocsai, P. and Shams, S., Visualization and analysis of gene expression data, *JALA*, 4, 58, 1999.

8 Zeroing in on Essential Gene Expression Data

Stefanie Fuhrman, Shoudan Liang, Xiling Wen, and Roland Somogyi

CONTENTS

8.1 INTRODUCTION

There was a time, not long ago, when measuring the expression of 10,000 genes was an impractical task. Then DNA microarrays[1] came along and changed everything. Microarrays and other technologies, such as large-scale RT-PCR (reverse-transcription polymerase chain reaction)[2] and SAGE (serial analysis of gene expression),[3] allow the assay of hundreds or thousands of genes at one time. Given these powerful technologies, it is now time to ask two fundamental questions: (1) How can we approach biomedical problems with large-scale gene expression assays? and (2) How can we extract the essential information from the resulting data? These are not trivial questions. The first question relates to experimental design, and the second relates to data analysis.

Until the present, most researchers have approached their work by focusing on one gene at a time, with little need for data analysis. The use of microarrays will therefore involve a substantial shift in strategy. While the logic of experimental design will, of course, remain unchanged, microarrays encourage the measurement of gene expression across multiple conditions, such as time points, drug treatments, or anatomical regions. By introducing this extra dimension of multiple conditions, the value of large-scale expression data is vastly increased. In that context, it is now possible to ask questions about global gene interactions and compare genes based

0-8493-2285-5/02/$0.00+$1.50

on the complexity of their expression patterns. As explained later, these questions are related to basic biomedical problems such as drug target and toxicity marker discovery, and the development of new therapies for degenerative diseases.

Beyond experimental design and data collection, it is necessary to organize the data into a comprehensible form, and this leads to the subject of data analysis. For our purposes, data analysis is simply a way of prioritizing genes and arranging the data so that interesting patterns become obvious. On a smaller scale of, for example, 100 genes, some analyses could be performed by visual inspection of the data. However, this becomes impractical on the large scale of microarray data, for which computational techniques are required. We discuss, in nontechnical fashion, a number of published techniques useful in the analysis of microarray data, including the application of Shannon entropy, expectation ratio likelihood (ERL), clustering, and reverse-engineering. These techniques complement one another, with each one addressing a different aspect of the data. Although some of the data presented was generated with RT-PCR, the same principles of experimental design and data analysis apply to microarrays.

Efficient experimental designs for microarrays, coupled with appropriate data analysis methods, should lead to a significantly greater understanding of issues related to tissue development. A great deal of work remains to be done in areas such as normal development and aging, degenerative diseases, and drug responses. Microarrays provide a way of collecting sufficient information to address previously unapproachable problems in these areas.

8.2 SOLVING BIOMEDICAL PROBLEMS WITH MICROARRAYS

Microarrays offer the opportunity to study changes or differences in tissues on a global scale. The benefit of this is that it allows one to begin to construct gene interaction diagrams. Another name for this is *reverse-engineering*,[4–7] a step toward discovering how to regenerate injured or diseased tissue. Applications would include the regeneration of injured brains and spinal cords, and perhaps entire limbs. This possibility is not far-fetched considering that human tissues are known to possess some capacity for regeneration, even in the central nervous system. Further, animals have a capacity for regenerating tissues that normally do not regenerate significantly in humans (e.g., the rat spinal cord). By using the appropriate animal models and microarray data from thousands of genes across multiple time points and conditions, we should be able to determine the keys to inducing regeneration. Later, we will discuss more specifics about reverse-engineering.

Although reverse-engineering will probably take years to accomplish, there are other problems that can be approached more easily. These include the assignment of functions to new genes and drug target discovery. Computational methods capable of addressing these problems are now in use, and include clustering, Shannon entropy, and expectation ratio likelihood (ERL). By grouping (clustering) genes according to similarities in their expression patterns over time, we can hypothesize common functions for genes in the same cluster.[8–13] Shannon entropy[14] can be used for selecting the most interesting genes from among thousands assayed;[15] these select genes can then be considered as possible drug targets or toxicity markers,[15-17] depending on the purpose of the study. ERL[16] is useful in determining which genes have expression patterns significantly different from the corresponding patterns in the control samples. These techniques provide a method for zeroing in on the most informative data from microarray experiments, and should make the drug discovery process more efficient.

The most interesting genes from analyses of microarray data may include those previously unsuspected of involvement in the physiological processes being studied. This is an important benefit of large surveys. Beyond this, PCR can be used as a more sensitive measure to study the expression patterns of genes selected by computational analysis.

8.3 SELECTING DRUG TARGET CANDIDATES

Of the estimated 100,000 genes in the human genome, we can reasonably expect that only a small fraction will be involved in any single physiological process. Traditionally, any statistically significant difference between control and experimental samples has been considered an indication of a gene's involvement in the process being studied. By that measure, only a minority of genes assayed by microarrays are relevant in animal disease and injury models (see Figure 8.1). This circumstance can only benefit biomedical science by reducing the number of genes requiring further investigation.

To obtain a better understanding of disease and injury, however, it will be necessary to go beyond a simple "control vs. treated" or "n-fold-change" approach, and employ a *time series* of gene expression. The reason for this is simply that any process is a series of events; and by examining only a single time point, we will miss most of the events that comprise the process. Some examples of time series that are worthy of investigation include injury, regeneration, degeneration, apoptosis, differentiation, aging, development, responses to toxins, and learning and memory formation. With a time series, we can continue to use a significant difference between the control and experimental samples as a criterion for a gene's importance. However, Fuhrman et al.[15] have proposed a more efficient method: the application of Shannon entropy.[14]

Shannon entropy is a basic measure of information content. For our purposes, information content can also be described as the variability, complexity, or change in a gene expression pattern. The greater the variability of gene expression, the higher the entropy or information content. This concept is diagrammed in Figure 8.2. Because *change* in expression is the criterion used to determine whether a gene should be studied as a putative drug target, it has been proposed that high-entropy genes make good drug target candidates.[15] In contrast, a gene with zero entropy shows no variability or change in its expression, and would probably not be a particularly interesting subject for further investigation. Beyond this, unlike N-fold change, entropy takes into account an entire time series, anatomical pattern, or series of conditions, thereby taking fuller advantage of the available data.

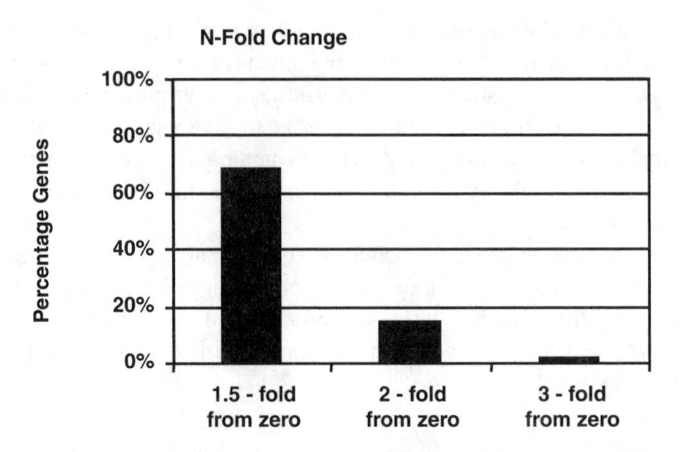

FIGURE 8.1 N-fold change as a measure of gene expression analysis in a study of the effects of a toxic dose of benzo(*a*)pyrene on rat liver (data provided by Mary Jane Cunningham). Of 4000 functionally annotated genes, only 15% show at least a twofold upregulation or downregulation by 1/2 from the control (vehicle-injected) zero time point ("2-fold from zero"). Only 3% exhibit at least a threefold upregulation or downregulation by 2/3 ("3-fold from zero"). These results are similar to those in other unpublished studies. In general, significantly up- or downregulated genes are only a small percentage of the total assayed given the administration of a drug or toxin. A twofold change from the control is generally considered a significant response for microarray assays, given their sensitivity, but this remains to be studied further. Here, "1.5-fold from zero" refers to a 1.5-fold upregulation or a 0.33 downregulation.

The use of Shannon entropy as a measure of anatomical complexity may be particularly useful in neuroscience, in which researchers must consider the distribution of gene expression across brain regions. We can simply substitute anatomical regions, such as brain nuclei, for the time points shown in Figure 8.2, and calculate the entropy. In this case, genes with high entropy can be regarded as those most responsible for anatomical complexity at a particular time point. It should be stressed, however, that time series remain an important component of such an analysis. For example, a gene with high entropy at a particular time point may have low entropy at some later time, especially if the brain is undergoing rapid change, such as degeneration or recovery from injury. Anatomical and temporal analyses are therefore best used in combination.

Shannon entropy can also be applied to a series of conditions. This involves a simple substitution of conditions for time points or anatomical regions. For example, entropy can be useful in identifying genes that distinguish among various tumor types. In this case, each tumor type is represented as an event (like a time point). Genes that show the highest entropy (greatest variability) across tumor types can be further investigated because these are the genes that make the tumor types unique. In general, a more varied set of tumor or tissue types should show higher entropy for the genes assayed, but this remains to be studied.

Potential problems with this application of entropy are explained by Fuhrman et al.[15] One of these is a possible source of false high-entropy values related to the

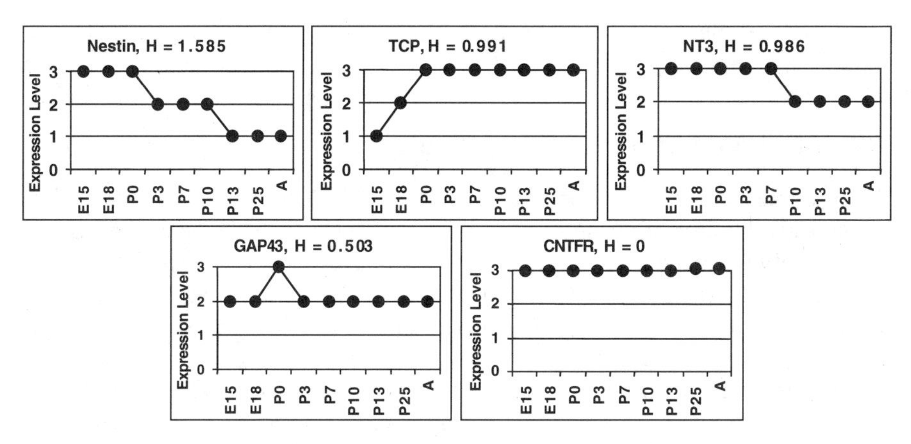

FIGURE 8.2 Shannon entropy for gene expression patterns from a study of rat hippocampus development. Temporal gene expression patterns are shown from a study of the developing rat hippocampus (unpublished) from embryonic day 15 (E15) through early postnatal stages (P0–25) to Adult (A, postnatal day 60). To calculate entropy, it is necessary to normalize the data to maximal expression so that the highest expression value over the time course for each gene is 1.0. We then bin the expression data into levels of expression. Eight or more time points permit three levels of expression (level 1, expression value is less than 0.33; level 2, expression is 0.33 or greater but less than 0.67; level 3 is greater than 0.67). Entropy is then calculated as $H = -\ p_i \log_2 p_i$, where p is the probability or frequency of occurrence of a level of gene expression *i*. Nestin has the highest possible entropy value for nine time points because it is expressed at all three levels with the least redundancy — three expression values at each of the three levels. The greater the redundancy of the pattern, the lower the entropy. TCP (T-complex protein) is also expressed at all three levels, but most of the pattern is at one level, making it more redundant than the nestin pattern. The most redundant pattern is that of CNTFR (ciliary neurotrophic factor receptor), which is invariant over time. The pattern for CNTFR contains no information and therefore has an entropy of zero. We could say that nestin has the most "active" pattern and is a major participant in hippocampal development. On the other hand, CNTFR is expressed at a constant level in the background and does not appear to be a significant effector of hippocampal development. Other abbreviations: NT3, neurotrophin 3; GAP43, growth-associated protein 43.

correlation of gene expression with protein expression. For example, a high-entropy gene expression pattern might be associated with a low-entropy pattern for expression of the corresponding protein product. These correlations remain to be studied on a large scale, but a parsimonious assumption is that proteins fluctuate in parallel with gene expression.

Entropy can be particularly useful for large microarrays containing thousands of genes, and can be applied easily using spreadsheet software.[17] Expression surveys of this magnitude require a simple method for organizing data and selecting the most relevant genes. For example, entropy can be used to rank-order a list of 8000 genes from an experiment involving a time course. It is then possible to focus further research efforts on genes at the top of the list. From that list, a researcher could then select genes from functional classes of interest, such as G protein-coupled receptors. In this way, thousands of potentially interesting genes can be narrowed down to a short, manageable list suitable for further investigation at the level of individual genes. The same concept can be applied to protein expression data.

8.4 CONTROL TIME SERIES AND ERL

In experiments involving time series, such as a study of an animal model of a degenerative disease, it is important to account for non-disease-related gene expression activity by including a control time series. In general, we can expect the control to exhibit little change in expression over time. However, there may be exceptions, especially if the model involves immature animals. In this situation, normal developmental events will be occurring in the background, and it will be important to separate these events from those associated with the pathology. One way of doing this is to use time point-matched controls in the microarray assays: each experimental time point sample is hybridized to the microarray along with a control from the same time point.

If, however, one wanted to observe normal development, the control time series would have to be assayed separately. In this case, each time point sample in the experimental series is hybridized to the microarray along with the first (i.e., zero) time point, so that gene expression at every time point is relative to the starting time. The same is done for the control series. This results in two separate time series of gene expression that must be compared. ERL (expectation ratio likelihood) has been proposed as a method of comparison for these time series.[16]

ERL compares time series based on both N-fold change and the shapes of the expression patterns over time. If the control and experimental temporal expression patterns are different, a gene will have a high ERL score. This is further explained in Figure 8.3.

ERL can be combined with Shannon entropy because the two measures complement each other. ERL addresses the problem of the control time series, selecting genes that exhibit a different expression pattern from the control. Shannon entropy can be applied either before or after ERL, to select the major participants in the process being studied. Some genes may have a high ERL score based primarily on absolute differences in gene expression between control and experimental time series; these would be eliminated by entropy, leaving only those genes that show variability in expression over time. Thus, by combining these measures, we may discover genes that are both specific for, and major effectors of, a physiological process.

8.5 CLUSTERING

With the sequencing of human, animal, and microbial genomes, we are faced with the problem of newly discovered genes with no known function. Until recently, molecular biologists have relied on nucleotide sequences for clues to gene functions, but these new sequences are not homologous to known genes. Today, we can use a different approach to this problem: clustering of gene expression patterns.[2,7–11,16,17] If a new sequence has an expression pattern similar to that of a familiar gene, we can hypothesize that the two genes have related functions. Hypotheses such as these will, of course, require testing on an individual basis.

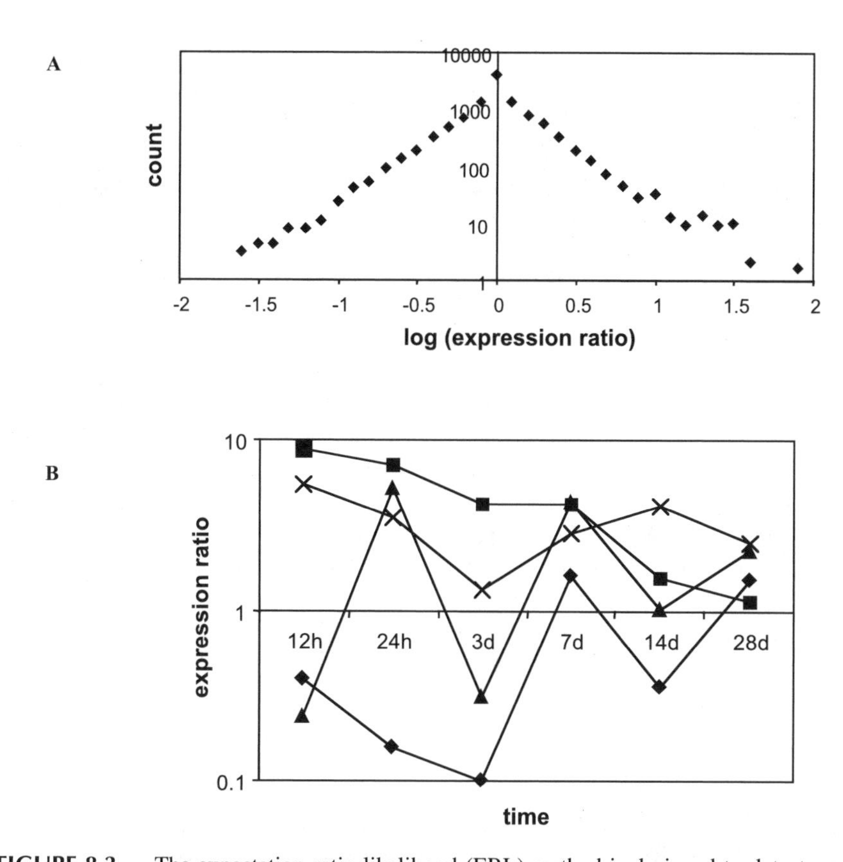

FIGURE 8.3 The expectation ratio likelihood (ERL) method is designed to detect genes whose expression in the drug-treated animals is significantly different from the normal course of development based on empirically determined p values. We set ERL = abs $(\log(R_t/R_c))$, where the R_t and R_c are expression ratios in the treated vs. control experiments; the sum is over the time course of the treated/control experiments. The higher the ERL value, the more unlikely that the expression difference is due to measurement errors and therefore more likely that the measured difference is a meaningful one. (A) The histogram of the logarithm of expression ratios for pairs of clones that independently measure the expression of the same gene on a microarray; 460 such pairs were used in constructing this histogram. The graph is based on 23 hybridizations (time points) from an experiment involving three drugs. This histogram indicates that the score function, which is defined as the negative logarithm of the expected random chance, should be proportional to the abs(log(expression ratio)), where abs( ) is the absolute value function. According to the histogram, most pairs of clones have a low log(expression ratio), indicating a low measurement error for the microarray assay. (B) Some typical time courses of genes with high ERL scores. Plotted are the ratios of drug-treated gene expression values to control values at 12 hours, 24 hours, 3 days, 7 days, 14 days, and 28 days after treatment. Squares and triangles are time courses for two genes responding to benzo(*a*)pyrene. Crosses and diamonds are the time courses for another two genes responding to acetaminophen. All the ERL scores are in the range between 6.5 and 7.5. From these four different patterns, we can see that high ERL score values can be due to consistent up- or downregulation of a gene (squares and crosses), or can be due to a very high expression value at a few time points or oscillating behavior (triangles and diamonds).

As with Shannon entropy, genes can be clustered according to expression patterns over time, anatomy, and conditions. Unlike entropy, clustering takes into account the shape of the expression pattern. An example of clustering is shown in Figure 8.4. Clustering is simply the grouping together of similar patterns. In the case of a relatively small number of genes, clustering can be done by eye, with the researcher determining which genes have similar patterns. For larger datasets, however, computational methods become necessary.

In Figure 8.4, we used FITCH[18] to cluster genes based on their temporal expression patterns. The first step in clustering is the generation of a distance matrix for all the genes. The distance matrix is similar to a milage chart, in which every gene is determined to be at some "distance" from every other gene on the list. The distances are determined by a correlation or distance measure, such as Pearson's correlation coefficient, euclidean distance, or mutual information. Euclidean distance, which captures positive, linear correlations, was used for the data in Figure 8.4. Genes with small euclidean distances between them will cluster together in the dendrogram created by the FITCH program (Figure 8.4).

Figure 8.5 shows a magnification of part of the dendrogram for the data in Figure 8.4. This branch of the dendrogram corresponds to "Wave 1" in Figure 8.4. Wave 1 contains two expressed sequence tags (ESTs) of unknown function (SC6 and SC7). Clues to the functions of these two genes can be found by observing

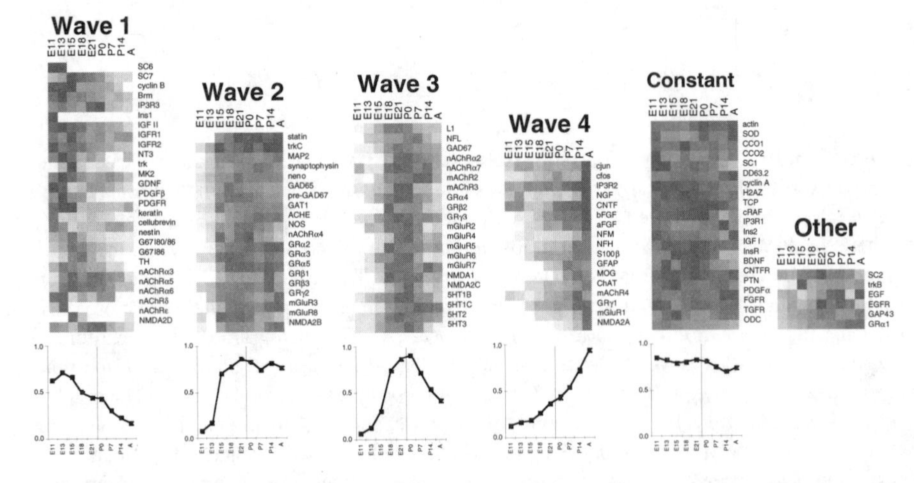

FIGURE 8.4 Clustered temporal gene expression patterns of rat spinal cord development. Genes were assayed using RT-PCR; 112 genes from different functional categories were assayed at nine developmental stages ranging from embryonic day 11 (E11) through postnatal ages 0–25 days (P0–25), and adult (A, 90 days postnatal). Gene expression was normalized to maximal expression, so the highest expression for each gene is 1.0. For each gene, the darkest color is the highest expression detected for the time series, and white means undetectable. Each "wave" of expression is considered a cluster. Note the GABA receptors (GR) clustered in wave 2 and the metabotropic glutamate receptors (mGluR) clustered in wave 3. (From Wen et al., *Proc. Natl. Acad. Sci. U.S.A.*, 95, 334, 1998. With permission.)

which genes cluster closely with them. In this case, both ESTs cluster with neurotransmitter-associated genes (acetylcholine and GABA related). We can therefore hypothesize that SC6 and SC7 are involved in cholinergic and GABAergic neurotransmission. The next step would be individual follow-up experiments on SC6 and SC7; antisense, for example, could be used to determine whether there are any effects on cholinergic or GABAergic transmission. Microarrays can also be used in these follow-up studies to check for effects on genes related to neurotransmission, such as acetylcholinesterase, GABA receptors, etc.

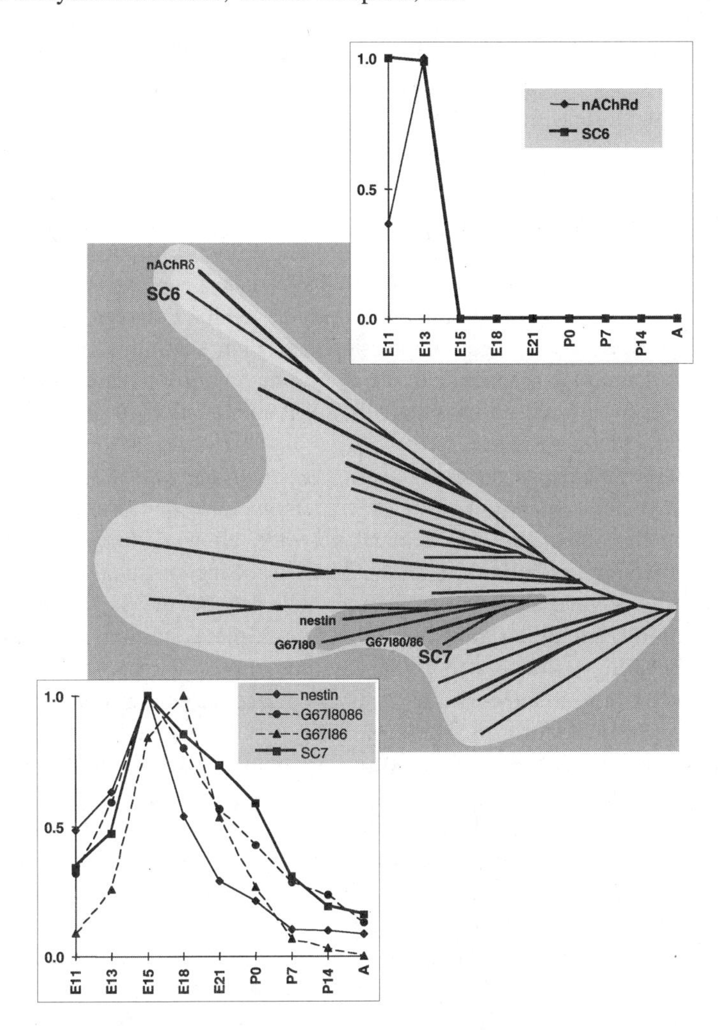

FIGURE 8.5 Enlargement of a branch of the cluster dendrogram that is the basis for Figure 8.4. The EST (SC6) clusters with a nicotinic acetylcholine receptor gene, while SC7 clusters with nestin and two splice variants of glutamate decarboxylase. Clustering can be used to suggest functions for new genes because genes that have similar expression patterns may function in related roles.

In addition, we should observe familiar genes with shared functions clustering closely together. This is certainly the case for GABA-A receptors (GR) and for PDGF-beta (PDGFb) and its receptor (PDGFR) (see Figure 8.4). In large-scale gene expression studies, we can therefore observe relationships consistent with traditional single-gene studies, as well as new, unexpected associations that should advance our understanding of the genetic network.

Another possible application for clustering lies in identifying drug candidates for combinatorial therapies. It may be advantageous in some cases to apply drugs in combination, and clustering can provide clues as to which genes might function together as drug targets. Shannon entropy might be helpful in this application as well. We can select the genes with high entropy in an animal model of disease or injury, thereby clustering only those genes that are the biggest participants in the disease process.

8.6 REVERSE-ENGINEERING

The most ambitious long-term objective of large-scale gene expression studies is reverse-engineering — the determination of an interaction diagram for all genes in the genetic network based on gene expression data (Figure 8.6). It should be possible to use such a diagram to regenerate tissue by turning on or off the appropriate genes. This will probably involve the manipulation of a combination of genes, in parallel and in series.

Some algorithms for reverse-engineering a genetic network have already been published.[4–6] In addition, Arkin et al.[19] have demonstrated that reverse-engineering can be applied to an actual biochemical network, allowing for some degree of accuracy in making predictions about the behavior of the chemical reactions.

Generating a gene interaction diagram will require the same type of data described above, including time series and multiple conditions. For example, reverse-engineering a tissue will require a time series of responses to multiple perturbations or drug treatments. The number of necessary drug treatments (conditions) has not been determined. However, Liang et al.[5] have demonstrated, using a computer model, that it takes about 100 perturbations to reverse-engineer a 50-gene binary network. This may be a surprisingly tiny number given that such a network contains 2^{50} (approx. 10^{15}) possible patterns for 50 genes, each of which can be either on or off. The number of perturbations needed to reverse-engineer a network with 10,000 or more genes remains to be determined, but even 100 perturbations will involve considerable laboratory work.

An efficient way of approaching the problem of perturbations is the application of a wide variety of drug treatments. High variability among perturbations should generate high variability in the resulting gene expression patterns. In contrast, the use of only one or a small number of drug classes should produce a very restricted range of tissue responses. This is related to the concept of Shannon entropy discussed earlier: greater variability means greater information content. The idea here is to extract as much information as possible using the fewest perturbations. Strategies such as this will be important in making reverse-engineering practicable.

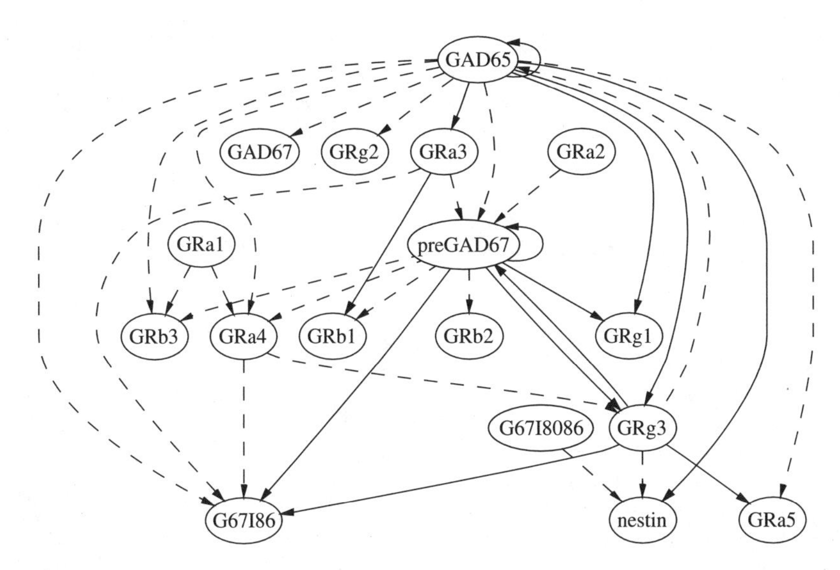

FIGURE 8.6 Reverse-engineering. This is a subset of a gene interaction diagram based on data from spinal cord[2] and hippocampus (unpublished results). Reverse-engineering requires both time series and multiple conditions. In this case, D'haeseleer et al.[6] have used data from three experiments (three conditions: spinal cord development, hippocampal development, and hippocampal seizure injury) and a linear model of gene interactions to generate an interaction diagram for 65 genes. The sub-network shown here contains proposed interactions among genes related to GABA-ergic neurotransmission. Solid lines refer to upregulation, and dashed lines to downregulation. (From D'haeseleer et al., *Proc. Pacific Symp. Biocomputing '99*, 41–52. With permission.)

8.7 CONCLUSIONS

New data analysis methods will be critical in making microarray and other large-scale gene expression data intelligible. Models of degenerative diseases and recovery from injury are particularly well suited to the type of experiments and analysis described here because they involve time series of changes in gene expression. With the appropriate experimental designs and data analysis techniques aimed at specific biomedical problems, it should be possible to make significant advances in developing therapies for degenerative diseases such as Alzheimer's and Parkinson's disease, and for spinal cord and brain injuries caused by trauma or stroke.

REFERENCES

1. Schena, M., Shalon, D., Davis, R.W., and Brown, P.O., Quantitative monitoring of gene expression patterns with a complementary DNA microarray, *Science*, 270(5235), 467, 1995.
2. Wen, X., Fuhrman, S., Michaels, G.S., Carr, D.B., Smith, S., Barker, J.L., and Somogyi, R., Large-scale temporal gene expression mapping of central nervous system development, *Proc. Natl. Acad. Sci. U.S.A.*, 95, 334, 1998.

3. Velculescu, V.E., Zhang, L., Vogelstein, B., and Kinzler, K.W., Serial analysis of gene expression, *Science*, 270(5235), 484, 1995.
4. Somogyi, R., Fuhrman, S., Askenazi, M., and Wuensche, A., The gene expression matrix: towards the extraction of genetic network architectures, *Proc. Second World Congress of Nonlinear Analysts (WCNA96)*, Elsevier Science, 1997, 1815.
5. Liang, S., Fuhrman, S., and Somogyi, R., REVEAL, a general reverse engineering algorithm for inference of genetic network architectures, *Proc. Pacific Symp. Biocomputing '98 (PSB98)*, World Scientific Publishing Co. Pte. Ltd., Singapore, 1998, 18.
6. D'haeseleer, P., Wen, X., Fuhrman, S., and Somogyi, R., Linear modeling of mRNA expression levels during CNS development and injury, *Proc. Pacific Symp. Biocomputing '99 (PSB99)*, World Scientific Publishing Co. Pte. Ltd., Singapore, 1999, 41.
7. D'haeseleer, P., Liang, S., and Somogyi, R., Genetic network inference: from co-expression clustering to reverse engineering, *Bioinformatics*, 16(8), 707–726, 2000.
8. Michaels, G.S., Carr, D.B., Wen, X., Fuhrman, S., Askenazi, M., and Somogyi, R., Cluster analysis and data visualization of large-scale gene expression data, *Proc. Pacific Symp. Biocomputing '98 (PSB98)*, World Scientific Publishing Co. Pte. Ltd., Singapore, 1998, 42.
9. Carr, D.B., Somogyi, R., and Michaels, G., Templates for looking at gene expression clustering, *Statistical Computing and Graphics Newslett.*, 8(1), 20, 1997.
10. Eisen, M.B., Spellman, P.T., Brown, P.O., and Botstein, D., Cluster analysis and display of genome-wide expression patterns, *Proc. Natl. Acad. Sci. U.S.A.*, 95, 14863, 1998.
11. Heyer, L.J., Kruglyak, S., and Yooseph, S., Exploring expression data: identification and analysis of coexpressed genes, *Genome Res.*, 9, 1106, 1999.
12. Walker, M.G., Volkmuth, W., Sprinzak, E., Hodgson, D., and Klingler, T., Prediction of gene function by genome-scale expression analysis: prostate-cancer associated genes, *Genome Res.*, 9, 1198, 1999.
13. White, K.P., Rifkin, S.A., Hurban, P., and Hogness, D.S., Microarray analysis of *Drosophila* development during metamorphosis, *Science*, 286, 2179, 1999.
14. Shannon, C.E. and Weaver, W., *The Mathematical Theory of Communication*, University of Illinois Press, Champaign, IL, 1963.
15. Fuhrman, S., Cunningham, M.J., Wen, X., Zweiger, G., Seilhamer, J.J., and Somogyi, R., The application of Shannon entropy in the identification of putative drug targets, *Biosystems*, 55, 5, 2000.
16. Cunningham, M.J., Liang, S., Fuhrman, S., Seilhamer, J.J., and Somogyi, R., Gene expression microarray data analysis for toxicology profiling, *Ann. N.Y. Acad. Sci.*, 919, 52–67, 2000.
17. Fuhrman, S., Cunningham, M.J., Liang, S., Wen, X., and Somogyi, R., Making sense of large-scale gene expression data with simple computational techniques, *Am. Biotechnol. Lab.*, (in press).
18. Felsenstein, J., PHYLIP (Phylogeny Inference Package) version 3.5c, Department of Genetics, University of Washington, Seattle, 1993.
19. Arkin, A., Peidong, S., and Ross, J., A test case of correlation metric construction of a reaction pathway from measurements, *Science*, 277, 1275–1279, 1997.

9 Application of Arrayed Libraries for Analysis of Differential Gene Expression Following Chronic Cannabinoid Exposure

Josef Kittler, Shou-Yuan Zhuang, Chris Clayton, Don Wallace, Sam A. Deadwyler, and Elena V. Grigorenko

CONTENTS

0-8493-2285-5/02/$0.00+$1.50

9.1 INTRODUCTION

In the past several years, a new technology called DNA array has attracted tremendous interest among biologists. DNA arrays, macro and micro formats, present large sets of DNA sequences immobilized onto solid substrates, glass slides, or nylon membranes. Based on hybridization methods that had been used for decades to identify and quantitate the expression of single gene (e.g., Southern and Northern blots, slot blots, library screening, etc.), and with help from high-speed robotics, DNA array technology can be applied to different tasks, including genetic mapping studies, mutational analyses, and large, genome-based scale monitoring of gene expression or identification of novel genes.

Arrayed cDNA libraries play an increasignly important role in the identification of specific gene expression profiles because they effectively represent the starting mRNA population. In arrayed libraries, each clone is picked and stored independently in separate 96- or 384-well microtiter plates. Thus, libraries can be quickly and accurately replicated and inoculated using various robotic devices and clone inserts can be amplified in the high-throughput PCR and used for construction of high-density grids.

9.2 AVAILABILITY OF ARRAYED LIBRARIES

A large number of arrayed cDNA libraries are available from different sources — both government and commercial — and each year more libraries are being added to the existing ones. The major distributors of high-density grids include the American Tissue Type Culture Collection (Rockville, MD), Genome Systems (St. Louis, MO), Research Genetics (Huntsville, AL), OriGene Technologies (Rockville, MD), Stratagene (La Jolla, CA), the UK HGMP Resource Center (Cambridge, U.K.) and the Resource Center and Primary Database (Berlin, Germany). All seven have appropriate information regarding the species and tissue source on the Web (<www.atcc.org>, <www.genomesystems.com>, <www.resgen.com>, <www.origene.com>, <www.stratagene.com>, <www.hgmp.mrc.ac.uk>, and <www.rzpd.de>, respectively).

9.3 HARDWARE FOR MAKING PRINTED ARRAYED LIBRARIES

Arraying of libraries with a small number of clones (<5000) can be carried out manually using sterile toothpicks. However, picking libraries with a large number of clones (>10,000) or several libraries from different tissue/cell sources cannot be achieved without robotic systems. Various robotic systems are now commercially available for this purpose and some equipment born in the laboratories has been described in detail on the Web (http://genome_www.stanford.edu/hardware). Usually, DNA clones are picked with 96-pin picking device and then transferred to a 96- or 384-well microtiter plate. A charge-coupled device (CCD) camera monitors the images of plates and specialized image analysis software calculates the position of clones. More than 3000/hour can be routinely achieved using this procedure.

9.4 AMPLIFICATION BY POLYMERASE CHAIN REACTION (PCR) AND GRIDDING OF LIBRARY INSERTS

The next step in the production of library arrays is the PCR amplification of cDNA inserts. For PCR amplification, cDNA clones that have been picked and grown in microtiter plates can be amplified by transferring a small amount of culture, using a 384-pin device, onto plates prefilled with a PCR mixture containing *Taq* polymerase (Perkin-Elmer, Palo Alto, CA), *Pfu* polymerase (Stratagene, La Jolla, CA), and vector-specific primers in 50 μl containing 0.2 m*M* dNTPs and 1× Promega PCR buffer. It is important to limit the PCR mixture to template primers, nucleotides, magnesium, reaction buffer, and enzyme. Additives such as glycerol and gelatin can interfere with the printing process by altering the surface tension of the drop. In the experiments described below, the PCR conditions were 95°C for 1 min; 94°C for 30 s, 50°C for 1 min, and 72°C for 2 min for 30 cycles; and 72°C for 5 min as a final elongation step. The plates are then heat-sealed using a commercial plate-sealing device (Genetix, U.K.).

Using these experimental conditions, agarose gel electrophoresis revealed the average cDNA insert size to be 1.5 kilobases, with an 80% success rate for PCR amplification. The addition of a proofreading DNA polymerase (such as *Pfu*) increased both the success rate and yield, especially for longer products. Because we used the cDNA library that was cloned into pSport1 vector, the M13 forward (a 32mer: 5′ GCTATTACGCCAGCTGGCGAAAGGGGGATGTG 3′) and M13 reverse (a 32mer: 5″CCCCAGGCTTTACACTTTATGCTTCCGGCTCG3′) primers were utilized in the PCR reaction.

The PCR products are gridded in duplicate onto positively charged 10 × 12-cm nylon filters (Boehringer Mannheim, Indianapolis, IN) using an automated 384-pin gridding device (Genetix Q-bot, U.K.). Each 10 × 12-cm cDNA array contained 12,228 different clones in duplicate; hence, each probe was assessed on two arrays comprising a total of 24,456 different cDNA clones (Figures 9.1 and 9.2). The filters are then denatured (0.5 *M* NaOH, 100 m*M* NaCl) for 2 min, neutralized (1 *M* Tris, pH 7.2; 100 m*M* NaCl) for 2 min, and UV-crosslinked in Stratalinker (Stratagene, La Jolla, CA).

To determine the variability of PCR and robotic arraying of PCR products, a series of hybridizations was performed on the arrays using an M13 promoter oligonucleotide probe internal to the PCR products. There was less than 5% variation between arrays in the amount of template DNA spotted at a particular coordinate (Figure 9.3A, top panel) in the arrays used to compare the hybridization profiles of labeled probes from either Δ^9-THC and vehicle-injected animals (Figure 9.3A, bottom panel). Whereas the M13 promoter oligonucleotide probe labeled all 48,912 spots on the two arrays, probes derived from hippocampal tissue hybridized to an average of only 11,225 of these spots, including duplicates, or roughly 23% of the total number of spots in the grid. The fact that hippocampal probes hybridized to only this many locations on the array was expected because the RNA used to produce the cDNA probes most likely did not contain all the genes represented in the whole rat brain cDNA library. This was verified by the fact that cDNA probes derived from

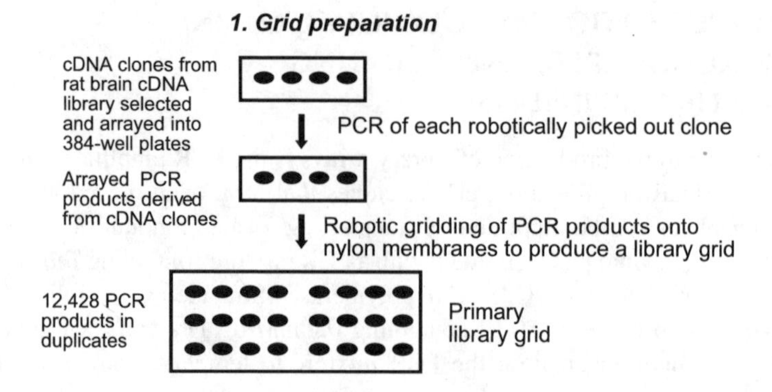

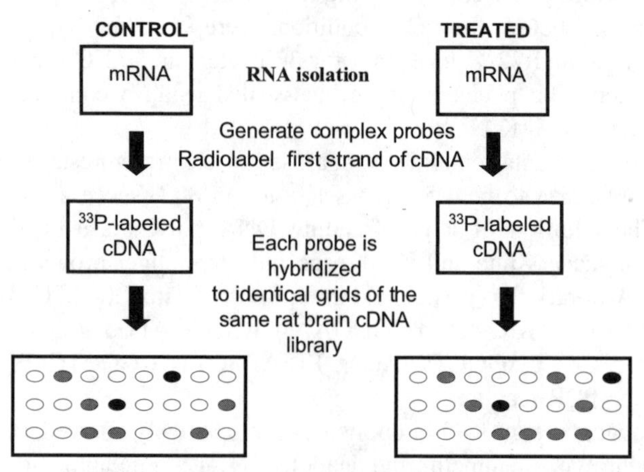

FIGURE 9.1 Schematic of application of arrayed cDNA libraries for analysis of differential gene expression. The diagram shows the method of arraying and assessing gene expression differences in RNA isolated from control and treated tissues.

RNA from different sources in the same animals produced a different hybridization pattern.

9.5 COMPLEX PROBE PREPARATION AND HYBRIDIZATION

Radiolabeled probes are prepared using 4 μl ^{33}P-dCTP (1-3000 Ci/mmol, Amersham, Piscataway, NJ) and 1 μg poly (A) RNA, or 20 μg of total RNA by anchor oligo dT primed first-strand synthesis using Superscript II reverse transcriptase (Gibco BRL, Gaithersburg, MD). Probes could be purified using Sephadex G-50 spin columns according to manufacturer instructions (Pharmacia Biotech, Piscataway, NJ). The quality and length of complex probes are checked by polyacrylamide gel electrophoresis on a 6% urea-tris-borate-EDTA (0.045 *M* tris-borate, 0.001 *M* EDTA,

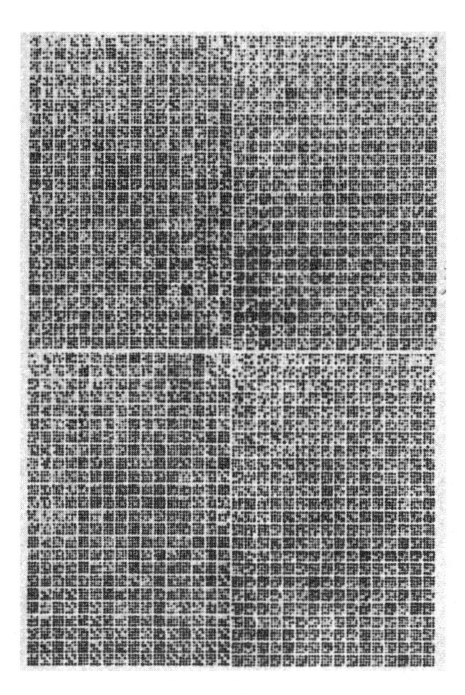

FIGURE 9.2 High-density rat brain cDNA library array. Each 10 × 12-cm cDNA nylon array contained 12,228 different clones in duplicate, comprising a total of 24,456 spots on each array. Probed with M13 oligonucleotide.

pH 8.0) buffered polyacrylamide gel. cDNA arrays are usually pre-hybridized in 15 ml DIG Easyhyb (Boehringer Mannheim, Indianapolis) at 45°C for 30 min in a rotisserie oven. Then, 10 μg human COT1 DNA (Gibco BRL, Gaithersburg, MD) and 435 μl EasyHyb were added to each probe prior to denaturation at 100°C for 5 min. The probe solution was quenched at 45°C for 90 min and hybridized to gridded cDNA arrays for 72 h at 45°C in a total volume of 10 ml EasyHyb. After hybridization, three washes of 20-min duration at high stringency in saline-sodium citrate buffer consisting of 3 *M* NaCl, 0.5 *M* sodium citrate, pH 7.2, and 0.1% SDS, diluted to 1:200 in H_2O, were performed at 68°C. Filters were exposed to a phosphor storage screen for 72 h before image capture on a phosphoimaging device (Storm scanner, Molecular Dynamics, Sunnyvale, CA).

9.6 IMAGE ANALYSIS

Image analysis of cDNA array spot hybridizations could be performed using different commercially available software that can compare intensity values of identically located cDNA amplicon spots between two filters and normalized statistically using the median spot intensity of each filter for any differences in cDNA probe activity between filters. In our experiments, the selection criteria for differentially expressed clones were: (1) a minimum 1.5-fold difference in intensity for a particular cDNA spot comparison, and (2) consistency of intensity between duplicate cDNA clones

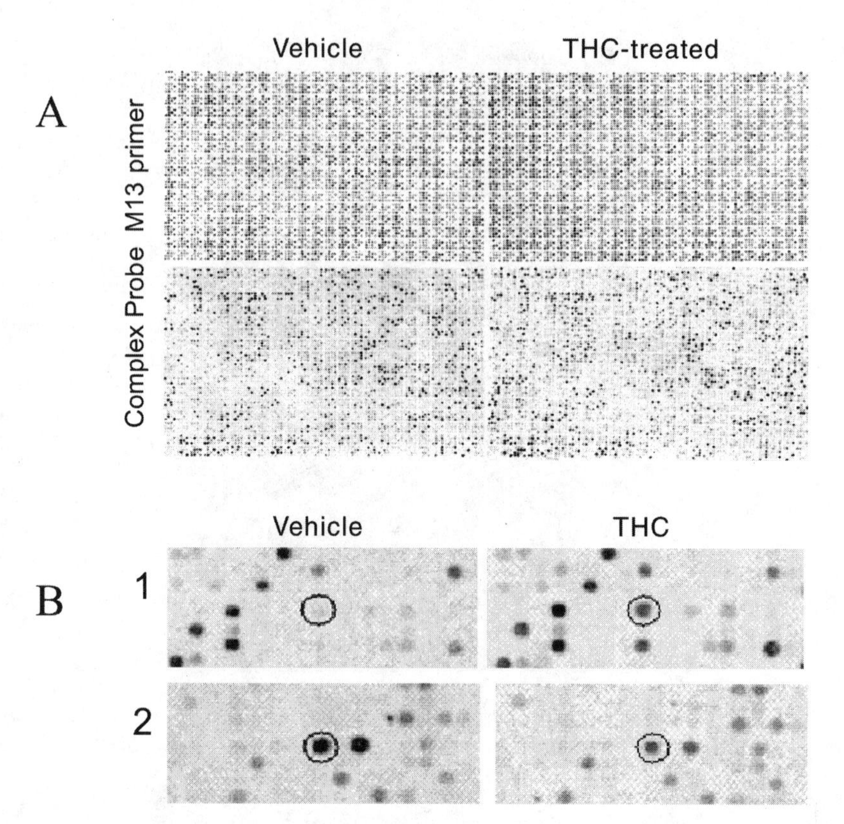

FIGURE 9.3 Enlargement of cDNA array image. (A) Each array shown in one fourth of the actual array. Each probe was assessed with two arrays containing different sets of 12,228 cDNAs in duplicate, for a total of 24,456 different cDNA clones per probe, or 48,912 total comparisons. Top: cDNA arrays hybridized with M13 promoter oligonucleotide internal to the PCR product for calibration and normalization purposes. Bottom: The same arrays were then hybridized to complex probes derived from hippocampus in vehicle and Δ^9-THC-injected (24 h) animals and compared for differences in "spot intensity" using DGENT software (GlaxoWellcome, Stevenage, U.K.). (B) Enlargement of the microarray array images at two different locations on the membrane and hybridized with complex cDNA probe from a vehicle-injected (Vehicle) and a Δ^9-THC (THC) injected (24 h) animal. The circled spots represent prostaglandin D synthase (upper panel #1) and myelin basic protein (lower panel #2) cDNAs. (From Kittler, J. et al., *Physiol. Genomics*, 3, 175–185, 2000. With permission.)

in the same array. Only candidate clones, which met these two criteria, were assessed by further sequencing and tests using RNA dot blot analyses.

9.7 DETECTION OF DIFFERENTIALLY EXPRESSED CLONES

We used an arrayed brain cDNA library for identification of altered gene expression in hippocampus following acute and chronic exposure to tetrahydrocannabinoid (Δ^9-THC), the primary psychoactive component of marijuana. Hybridization patterns of hippocampal mRNA from Δ^9-THC-exposed animals were compared to the patterns derived from vehicle-injected control animals (see Figure 9.3). The similarity of the overall hybridization patterns for probes from vehicle and Δ^9-THC-injected animals is quite apparent in Figure 9.3 because the majority of the 5605 genes expressed in hippocampus (23% of 24,456) were unchanged by acute or chronic Δ^9-THC exposure.

For each duration of Δ^9-THC exposure (24 h, 7 and 21 days), two different arrays containing 12,228 clones in duplicate were constructed and compared for each animal. The resolution of the technique only allowed genes with a ≥1.5-fold difference in labeled ^{33}P spot intensity to be considered significantly altered (see Table 9.1) and automatically excluded genes that did not show this degree of change in expression level.[1]

Comparative analyses of the hybridization patterns following different stages of Δ^9-THC exposure revealed a total of 180 detections of differential spot intensity across 24 h, 7 days, and 21 days of Δ^9-THC exposure. Sequence analyses revealed 159 clones representing known genes, 10 clones with high homology to sequences in the expressed sequence tag database (EST), and 11 clones without homology to any sequences in GenBank. Because a non-normalized rat whole brain cDNA library was employed, it was possible to have multiple detections of the same clones, indicated by the numbers in brackets in Table 9.1. Taking into account this redundancy, a net total 28 genes in hippocampus exhibited altered expression at various time points following acute or chronic exposure to Δ^9-THC. Table 9.1 shows the GenBank ID number of the 28 genes (and one EST), together with the closest homology match using an expected value (E) cutoff of $\leq 1.0\ e^{-20}$.[2]

Identified genes whose expression was altered by Δ^9-THC exposure were grouped by the following classification scheme: metabolism, cell adhesion/cytoskeletal proteins, myelination/glial differentiation, signal transduction proteins, and proteins involved in folding/or cell degradation processes. Clones that could not be placed in the above scheme were designated as "other" in Table 9.1.

A single, high (10 mg/kg) acute dose of Δ^9-THC initiated changes in expression within 24 h in 15 of the 28 altered genes that were dispersed across the different classifications is shown in Table 9.1. Six genes had transiently altered expression levels after the single dose of Δ^9-THC; four were upregulated (neural cell adhesion molecule [NCAM]), prostaglandin D synthase, 14-3-3 protein, and ubiquitin-conjugating enzyme); and two were downregulated (GDP dissociation inhibitor protein and transferrin). Changes in expression of these six genes returned to vehicle-injected control levels 7 days after chronic Δ^9-THC exposure (see below). The acute changes likely reflect the direct physiological consequences of the large dose (10 mg/kg) of Δ^9-THC, which produces severe catalepsy, hypothermia, and other symptoms in rats for the initial 3 to 5 days of repeated injections.[3] Expression levels of four other genes — SC1 protein, cytochrome oxidase, myelin proteolipid protein, and T-cell receptor protein (Table 9.1), were also upregulated after the single injection of Δ^9-THC but, unlike the above genes, remained significantly elevated throughout the entire 21 days of Δ^9-THC exposure. Another gene elongation factor remained upregulated for 7 days, but returned to control levels after 21 days of exposure. Finally, expression levels of the calmodulin and polyubiquitin transcripts changed in a

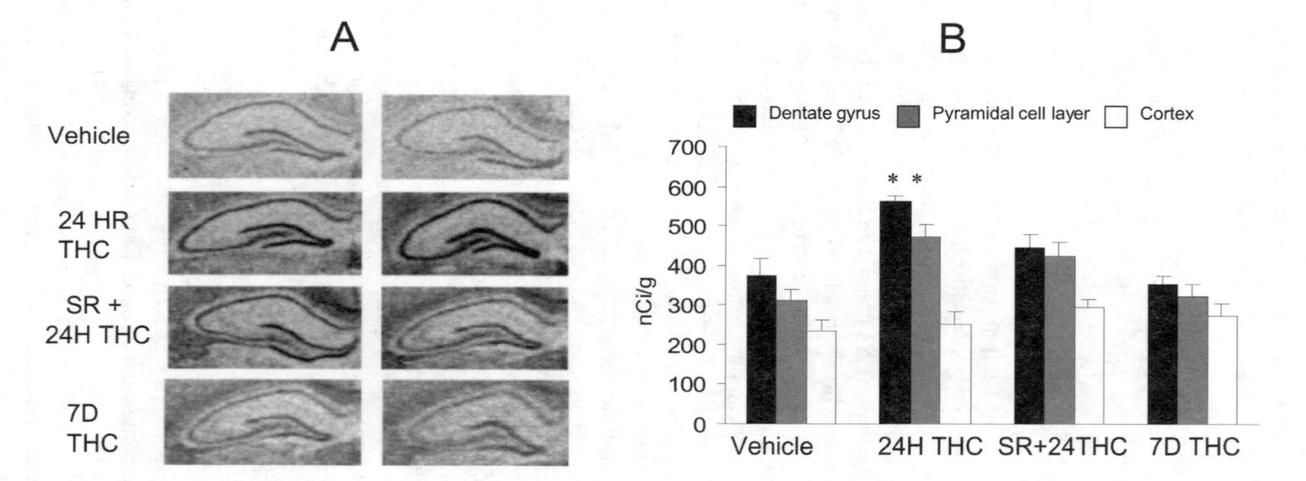

FIGURE 9.4 *In situ* autoradiograph of coronal sections of hippocampus from rats treated with vehicle, THC (24 h), SR171416A + THC (24 h), and THC (7 days). (A) NCAM labeling of hippocampal subregions in vehicle-, THC (24 h), SR + THC (24 h), and THC (7 days) treated animals. (B) Optical densities of dentate gyrus (black bars), pyramidal cell layer (gray bars) of hippocampus, and somatosensory cortex (white bars) were obtained from coronal sections and converted to cpm/mg tissue using a standard curve derived from brain paste (see methods). The values from three groups of animals (n = 3 in each group) are expressed as the mean ± SEM. The level of statistical significance was determined by ANOVA, $*p < 0.01$. (From Kittler, J. et al., *Physiol. Genomics*, 3, 175–185, 2000. With permission.)

biphasic manner across the duration of the chronic treatment period, as evidenced by downregulation after a single injection, reversing to upregulation at 7 days and, in the case of calmodulin, was again downregulated after 21 days (Table 9.1).

The overall pattern of gene expression, changing after 7 days of Δ^9-THC exposure, was different from that following a single acute injection. As indicated above, 8 of 15 genes whose expression was altered at 24 h remained altered at 7 days of Δ^9-THC exposure. In addition, transcripts for α-enolase, myelin-associated protein (upregulated), as well as phospholipid glutathione peroxidase, myelin basic protein, angiotensin AT1 receptor, G3PDH, and acidic 82-kDa protein (downregulated), were all affected after 7 days.

By day 21 of Δ^9-THC exposure, the expression levels of 16 of 28 genes either remained altered or became changed from their prior status at 7 days. At this time, seven genes were upregulated and 9 downregulated (Table 9.1) relative to vehicle-injected animals. Five of the nine downregulated genes (fructose-bisphosphate aldolase, brain lipid binding protein, proteosomal ATPase, calreticulin, and HSP70) were evident only after 21 days of exposure (Table 9.1), while expression levels of genes encoding myelin basic protein and mitochondrial 12-16S RNA , downregulated at 7 days, were also downregulated at 21 days (see Table 9.1). The upregulation in expression of calmodulin and elongation factor observed at 7 days was reversed to downregulation at 21 days of Δ^9-THC exposure (Table 9.1).

9.8 CONFIRMATION OF PRIMARY SCREEN

9.8.1 RNA Dot-Blot

To validate the specificity and magnitude of the Δ^9-THC-induced gene expression changes shown in Table 9.1, dot blots of mRNA analyses were employed to validate the expression patterns of 15 of the 28 genes identified in the primary screen. Table 9.2 shows the mean differences (±SEM) in expression levels for the selected genes examined in groups of three rats, each group exposed to Δ^9-THC for one of the three different durations (24 h, 7 days, and 21 days) relative to equal numbers of vehicle-injected controls.

The RNA dot blot analyses confirmed the changes in transcriptional regulation, which occurred at various stages of Δ^9-THC exposure indicated in the primary screen. In agreement with reports from other investigators, the magnitude of the differences in expression levels assessed by RNA dot blot analyses were, in general, lower relative to that reported in the primary screen.[4] There was considerable agreement, however, in the effect of duration of drug exposure on the direction (up- or downregulation) and relative magnitude of altered gene expression detected by both methods (Table 9.2).

9.8.2 *In situ* Hybridization

Increased expression of NCAM was detected and verified after a single dose of Δ^9-THC (10 mg/kg) in the cDNA microarray screen and RNA dot blot analyses (Tables 9.1 and 9.2). *In situ* hybridization techniques were therefore implemented

TABLE 9.1
List of Genes Differentially Expressed in Rat Hippocampus Following 24 h, 7 days, and 21 days of Δ^9-THC Exposure

Clone/Cellular Function	Expected Value (E)	GenBank ID Number	Hippocampus		
			24 h	7 day	21 day
		Metabolism			
Fructose-bisphosphate aldolase	7.1 e-107	M12919			↓
Glyceraldehyde-3-phosphate dehydrogenase	7.1 e-22	M17701		↓	
Cytochrome oxidase	5.0 e-171	S79304	↑ (15)	↑ (7)	↑ (43)
α-Enolase	2.9 e-59	X52379		↑	
Phospholipid glutathione peroxidase	8.2 e-55	X82679		↓	
		Cell Adhesion/Structural			
NCAM	3.1 e-124	X06564	↑		
SC1 protein	8.6 e-80	U27562	↑	↑	↑
Myelination/Glial Differentiation					
Myelin associated protein	4.5 e-77	X89638		↑	
Myelin proteolipid protein	8.5 e-116	M11185	↑	↑ (2)	↑
Myelin basic protein	8.2 e-137	M15060		↓ (2)	↓ (6)
Brain lipid binding protein	1.6 e-201	U02096			↓
Receptors/Transporters					
Angiotensin AT1 receptor	2.6 e-118	S66402		↓	↑ (12)
		Signal Transduction/Receptors			
Proteosomal ATPase	3.5 e-184	D83521			↓
Calmodulin	2.5 e-43	E02315	↓ (2)	↑ (2)	↓
Calreticulin	2.5 e-15	D78308			↓
Prostaglandin D synthase	4.9 e-147	J04488	↑ (11)		
14-3-3 Protein gamma subunit	4.2 e-51	D17447	↑		
GDP dissociation inhibitor	9.9 e-157	L07925	↓		
PKU beta subunit	9.4 e-52	AB004885			↑
		Protein Folding			
HSP 70	1.3 e-153	X70065			↓ (2)
Ubiquitin-conjugating enzyme	4.7 e-26	AF031141	↑		
Polyubiquitin	2.9 e-64	D17296	↓ (2)	↑	
Chaperonin containing TCP-1	6.2 e-115	Z31553	↑		
Others					
12S and 16S rRNA	1.1 e-132	J01438	↓	↓ (11)	↓ (7)
Acidic 82 kDa protein	2.9 e-59	U15552		↓	
Transferrin	7.1 e-119	D38380	↓		
Elongation factor	1.4 e-143	X63561	↑ (5)	↑ (2)	↓
T cell receptor beta locus	4.5 e-145	AE000663	↑	↑	↑

TABLE 9.1 (CONTINUED)
List of Genes Differentially Expressed in Rat Hippocampus Following 24 h, 7 days, and 21 days of Δ^9-THC Exposure

Clone/Cellular Function	Expected Value (E)	GenBank ID Number	Hippocampus 24 h	Hippocampus 7 day	Hippocampus 21 day
		EST •			
Clone 407167	5.5 e-91	AA048564			↑

Note: Expression level is measured as ratio of Δ^9-THC-treated vs. control samples for upregulated (↑) genes, and control vs. Δ^9-THC-treated samples for downregulated (↓) genes. Two array measurements were made per animal for each clone, for two different animals at each time point. Multiple detections of the same gene are in brackets. Matches of clone sequences from cDNA arrays were based on the calculated expect value (E), where $E \leq 1.0\ e^{-20}$ was considered as the criteria for sequence homology in the identified clones.

to reveal those regions of hippocampus in which the changes in NCAM expression occurred following acute (24-h) Δ^9-THC injections. Because the primary and secondary screens showed the change to be transient, returning to vehicle levels at 7 days, *in situ* histochemistry was also performed on tissue from animals after 7 days of exposure.

Figure 9.4 shows the distribution of NCAM in the dorsal hippocampus of rats treated with Δ^9-THC at 24 h and 7 days. It is clear that label was abundant in the hippocampal region, as revealed by an increased density in the pyramidal cell layers in Ammon's horn and granule cell layer in the dentate gyrus (DG) as previously reported.[5] Semiquantitative analyses of the density of label showed a significant increase (45%, $p < 0.01$) in density of the hybridization signal for NCAM-positive cells in the DG and pyramidal cell layers (42%, $p < 0.01$) 24 h after a single, acute injection of Δ^9-THC compared to vehicle-treated control animals (Figure 9.4B). The increase at 24 h was significantly reduced if the cannabinoid CB1 receptor antagonist SR171416A was injected 5 min prior to Δ^9-THC injection.[19] Consistent with results from the primary and secondary (dot blot) screens, after 7 days of chronic Δ^9-THC exposure, the *in situ* label increase in NCAM expression was no longer apparent. There were no significant changes in the density of NCAM *in situ* label in the somatosensory cortex in any of the animals that exhibited altered expression of NCAM in the hippocampus (Figure 9.4B).

9.9 CONCLUSIONS

An arrayed cDNA library was employed to examine alterations in gene expression in hippocampus that occurred as the result of both acute (24 h), intermediate (7 days), and long-term (21 days) exposure to Δ^9-THC. This large-scale gene expression analysis revealed 28 known genes, 10 previously sequenced ESTs, and 11 potential novel gene sequences that were altered at one or more of the above time points of

TABLE 9.2
Differential Gene Expression (DGE) in Rat Hippocampus Monitored by Primary and Secondary Microarray Screens and RNA Dot Blot Analyses

Clone	DGE Confirmed	Primary Screening (fold difference)			RNA Dot Blot (fold difference)		
		24 h	7 day	21 day	24 h	7 day	21 day
Prostaglandin D synthase	Y	1.8↑			1.6 ± 0.3↑*	1.2 ± 0.2↑	1.3 ± 0.2↑
Transferrin	Y	1.8↓			1.5 ± 0.2↓*	1.5 ± 0.1↓	1.2 ± 0.3↓
NCAM	Y	3.2↑			1.6 ± 0.2↑*	1.2 ± 0.3↑	1.1 ± 0.2↑
Sodium channel	N		1.9↑		1.6 ± 0.4↑	1.0 ± 0.1↑	1.3 ± 0.2↑
Enolase	Y		2.3↑		1.4 ± 0.1↑	1.9 ± 0.2↑*	0.9 ± 0.2↑
GST peroxidase	N		1.8↓		1.2 ± 0.1↓	1.3 ± 0.2↓	1.4 ± 0.2↓
PKU beta subunit	Y			2.5↑	1.0 ± 0.3↑	1.2 ± 0.4↑	1.7 ± 0.2↑*
Peptide-histidine transporter	Y			4.0↑	1.4 ± 0.3↑	1.0 ± 0.1↑	2.5 ± 0.4↑**
HSP70	Y			2.3↓	1.5 ± 0.1↑	1.6 ± 0.2↓	1.7 ± 0.3↓*
Proteosomal ATPase	Y			8.6↓	1.1 ± 0.05↓	0.8 ± 0.3↓	2.0 ± 0.5↓*
Brain-specific lipid binding protein	Y		2.5↓	2.5↓	1.2 ± 0.2↓	1.5 ± 0.1*↓	1.7 ± 0.2*↓
Angiotensin A1 receptor	Y/N		1.6↓	2.2↑	1.0 ± 0.3↑	1.2 ± 0.2↑	1.7 ± 0.2↑*
Myelin basic protein	Y		1.8↓	2.0↓	1.3 ± 0.4↓	1.5 ± 0.1↓*	1.6 ± 0.2↓*
SC1	Y/N	1.7↑	1.5↑	1.9↑	1.3 ± 0.2↑	2.0 ± 0.4↑*	2.7 ± 0.5↑**
Calmodulin	Y/N	1.6↓	2.2↑	1.5↓	1.5 ± 0.1*↓	1.8 ± 0.15↑*	1.4 ± 0.3 ↓

Note: Expression levels in RNA dot blot analyses were normalized to levels of labeled oligo(d)T primer for each blot (see methods) and expressed as a fold difference from corresponding vehicle-injected controls. Values for primary screens were determined from microarray scan software (Figure 9.2).

Values are mean ±SEM; (n = 3) $*p < 0.01$; $** p < 0.001$, *t*-test.

chronic Δ^9-THC exposure. Distinct patterns of differential transcript expression over the three stages of drug exposure were revealed in the primary screen (Tables 9.1 and 9.2). Verification of the findings agreed with other reports[4] regarding comparison with mRNA dot blots that gave reduced levels of change, but a high degree of concordance (70%) with the results from the large-scale primary screens. Therefore, the exceedingly small number of altered genes (n = 28, Table 9.1), identified by the primary and largely confirmed by the secondary (dot blot) screen (Tables 9.1 and 9.2), makes it unlikely that the changes reported here resulted from chance variations in expression levels of the identified clones. However, because proteins levels corresponding to the above mRNA changes were not monitored, the functional significance of the change in expression levels cannot be fully ascertained.

The altered expression level of several genes after a single injection of Δ^9-THC (Tables 9.1 and 9.2) could be produced by at least two (if not more) different cannabinoid receptor mediated cellular changes: (1) a decrease in the level of cAMP via receptor-coupled inhibition of adenylyl cyclase, which could eventually alter mRNA transcription through cAMP response element-binding protein (CREB) or other gene pathways;[5,6] or (2) the high dose (10 mg/kg) of Δ^9-THC may have acted via cannabinoid receptor-produced physiological changes such as severe catalepsy and hypothermia[7,8] to initiate expression of multiple transcripts, including immediate early genes.[9] Furthermore, genes identified (Table 9.1) and confirmed (Table 9.2) as differentially expressed after acute injection can be subdivided again into those affected only after a single injection, returning to vehicle-injected control levels after 7 or 21 days (i.e., NCAM, prostaglandin D synthase, transferrin), and those whose expression levels remained dysregulated over the entire 21 days of Δ^9-THC exposure (calmodulin, SC1). The transient increase in expression levels of NCAM only after a single acute dose of Δ^9-THC is interesting because it has been suggested that changes in cAMP levels lead to a downregulation of cell adhesion molecules involved in synaptic remodeling. Upregulation of NCAM expression has been traditionally associated with the maturation and differentiation of neurons in the developing rat hippocampus.[10] However, recently, a transient increase in NCAM expression has been shown in certain learning paradigms.[11] To date, there has been no report of NCAM's sensitivity to cannabinoid receptor activation nor that such changes are transient in nature and decrease with chronic exposure to Δ^9-THC. Our data suggests that acute Δ^9-THC treatment may alter NCAM levels in the hippocampus acutely as a consequence of LTP suppression, but following continued suppression of LTP by chronic Δ^9-THC treatment, NCAM expression is no longer affected.

The fact that the expression of calmodulin was bidirectionally altered at 24 h, 7 days, and 21 days may reflect similar phasic changes in cAMP levels following development of cannabinoid tolerance.[8] In contrast, the basis for the 21-day elevated level of SC1 (hevin), a newly cloned gene for a protein which binds to cytokine receptors, proteases, and matrix proteins,[12] throughout the 21 days of Δ^9-THC exposure is not well understood; however, the fact that its expression is changed so significantly suggests a possible convergence with cannabinoid receptor-mediated cellular changes.

Both the primary screen and RNA dot blot analyses showed significantly elevated prostaglandin D synthase expression after a single injection Δ^9-THC. Cannabinoids stimulate prostaglandin D synthetase activity[13] and the production of prostaglandins,[14] which is a likely the result of release of arachidonic acid, as has been shown in adult brain slices, neuroblastoma cells, and in cultured hippocampal neurons. Prostaglandin D2, the end product of this enzymes'reaction with arachidonic acid, is abundantly expressed in brain and has recently been shown to induce sleep in rats.[15] The above findings suggest that acute cannabinoid exposure can act directly to transiently increase expression of an enzyme present in one the synthesis pathways for endogenous cannabinoids.

By day 7 of drug exposure, animals do not exhibit catalepsy to the high (10 mg/kg) dose of Δ^9-THC, but do display normal locomotor behavior and remain significantly impaired in cognitive function.[1,16] Hippocampal CB1 receptor-stimulated GTPγS binding is maximally depressed at day 7 and coincides with peak elevation of CB1 receptor mRNA expression levels.[1] The increased expression of calmodulin (see above) at day 7 (Table 9.2) could therefore reflect increased levels of Ca^{2+}/calmodulin-stimulated adenylyl cyclase following CB1 receptor desensitization.[18]

Previous reports show that both behavioral tolerance as well as CB1 receptor downregulation and desensitization reach a steady state at 21 days of Δ^9-THC exposure,[18,19] at which point expression levels of the CB1 receptor are not different from vehicle control animals.[1]

Thus, application of brain arrayed library for analysis of gene expression at different stages of cannabinoid exposure successfully identified gene patterns within rat hippocampus that are specific for each stage. A number of transcripts involved in important cellular functions (i.e., NCAM, SC1, PKU beta subunit, HSP70, etc.) have altered expression levels at some or all time points during the extended course of Δ^9-THC treatment. Detection by differential gene expression criteria at these different time points (Tables 9.1 and 9.2) provided information about the complexity of cellular and molecular changes affected by both acute and chronic exposure to cannabinoids. Specifically, the technique revealed changes in several cellular processes not previously associated with the development of tolerance to cannabinoids.[20] Such changes in expression levels incorporate a wide variety of signaling mechanisms within hippocampal cells that were differentially susceptible to chronic cannabinoid treatment. The significance of many of these changes with respect to cannabinoid system function has yet to be determined. However, it is unlikely that such profound and widespread alterations in gene regulation revealed through cDNA microarray technology would not have an impact on hippocampal function.

REFERENCES

1. Zhuang, S.Y., Kittler, J., Grigorenko, E.V., Kirby, M.T., Hampson, R.E., and Deadwyler, S.A., Effects of long-term exposure to Δ^9-THC on the expression of CB1 receptor mRNA in different rat brain regions, *Mol. Brain Res.*, 62, 141–149, 1998.
2. Karlin, S. and Altschul, S.F., Applications and statistics for multiple high-scoring segments in molecular sequences, *Proc. Natl. Acad. Sci. U.S.A.*, 90, 5873–5877, 1993.
3. Adams, I.B. and Martin, B.R., Cannabis: pharmacology and toxicology in animals and humans, *Addiction*, 91, 1585–1614, 1996.

4. Patten, C., Clayton, C.L., Blakemore, S.J., Trower, M.K., Wallace, D.M., and Hagan, R.M., Identification of two novel genes by screening of a rat brain cDNA library, *Neuroreport*, 9, 3935–3941, 1998.
5. Carlezon, W.A., Jr., Thome, J., Olson, V.G., Lane-Ladd, S.B., Brodkin, E.S., Hiroi, N., Duman, R.S., Neve, R.L., and Nestler, E.J., Regulation of cocaine reward by CREB, *Science*, 282, 2272–2275, 1998.
6. Felder, C.C., Veluz, J.S., Williams, H.L., Briley, E.M., and Matsuda, L.A., Cannabinoid agonists stimulate both receptor- and non-receptor-mediated signal transduction pathways in cells transfected with and expressing cannabinoid receptor clones, *Mol. Pharmacol.*, 42, 838–845, 1992.
7. Fan, F., Compton, D.R., Ward, S., Melvin, L., and Martin, B.R., Development of cross-tolerance between Δ^9-tetrahydrocannabinol, CP 55,940 and WIN 55,212, *J. Pharmacol. Exp. Ther.*, 271, 1383–1390, 1994.
8. Hutchenson, D.M., Behavioral and biochemical evidence for signs of abstinence in mice chronically treated with delta-Δ^9tetrahydrocannabinol, *Br. J. Pharmacol.*, 125, 1567–1577, 1998.
9. Mailleux, P., Verslype, M., Preud'homme, X., and Vanderhaeghen, J.J., Activation of multiple transcription factor genes by tetrahydrocannabinol in rat forebrain, *Neuroreport*, 5, 1265–1268, 1994.
10. Seki, T. and Arai, Y., Distribution and possible roles of the highly polysialated neural cell adhesion molecule (NCAM-H) in the developing and adult central nervous system, *Neurosci. Res.*, 17, 265–290, 1993.
11. Murphy, K.J., O'Connell, A.W., and Regan, C.M., Repetitive and transient increases in hippocampal neural cell adhesion molecule polisialylation state following multitrial spatial training, *J. Neurochem.*, 67, 1268–1274, 1996.
12. Bornstein, P., Diversity of function is inherent in matricellular proteins: an appraisal of thrombospodin 1, *J. Cell Biol.*, 130, 503–506, 1995.
13. Reichman, M., Nen, W., and Hokin, L.E., Tetracannabinol increases arachidonic acid levels in guinea pig cerebral cortex slices, *Mol. Pharmacol.*, 34, 823–828, 1993.
14. Hunter, S.A. and Burstein, S.H., Receptor mediation in cannabinoid stimulated arachidonic acid mobilization and anandamide synthesis, *Prostaglandins Leukot. Essent. Fatty Acids*, 43, 185–190, 1991.
15. Urade, Y., Hayashi, O., Matsumura, H., and Watanabe, K., Molecular mechanism of sleep regulation by prostaglandin D2, *J. Lipid Mediat.*, 14, 71–82, 1996.
16. Aceto, M.D., Scates, S.M., Lowe, J.A., and Martin, B.R., Dependence on delta 9-tetrahydrocannabinol: studies on precipitated and abrupt withdrawal, *J. Pharmacol. Exp. Ther.*, 278, 1290–1295, 1996.
17. Deadwyler, S.A., Heyser, C.J., and Hampson, R.E., Complete adaptation to the memory disruptive effects of delta-9-THC following 35 days of exposure, *Neurosci. Res. Commun.*, 17, 9–18, 1995.
18. Sim, L.J., Hampson, R.E., Deadwyler, S.A., and Childers, S.R., Effects of chronic treatment with Δ^9-tetrahydrocannabinol on cannabinoid-stimulated [^{35}S]GTPS autoradiography in rat brain, *J. Neurosci.*, 16, 8057–8066, 1996.
19. Breivogel, C.S., Sim, L.J., Deadwyler, S.A., Hampson, R.A., and Childers, S.R., Chronic Δ^9-tetrahydrocannabinol treatment produces a time-dependent loss of cannabinoid receptor binding and activation of G-proteins in rat brain, *J. Neurochem.*, 73, 2447–2459, 1999.
20. Childers, S.R. and Deadwyler, S.A., Role of cyclic AMP in the actions of cannabinoid receptors, *Biochem. Pharmacol.*, 52, 819–827, 1996.

Index

A

B

C

D

H

I

K

L

M

N

Q

R

S

T

U

V

W

X

Y

Z